STRUCTURE AND BONDING

Volume 31

Editors:
J.D.Dunitz, Zürich · P.Hemmerich, Konstanz
J.A.Ibers, Evanston · C.K.Jørgensen, Genève · J.B.Neilands,
Berkeley · D.Reinen, Marburg · R.J.P.Williams, Oxford

With 40 Figures and 6 Tables

Springer-Verlag
Berlin Heidelberg GmbH 1976

ISBN 978-3-662-15504-2   ISBN 978-3-540-37997-3 (eBook)
DOI 10.1007/978-3-540-37997-3

Library of Congress Catalog Card Number 67-11280

© by Springer-Verlag Berlin Heidelberg 1976
Originally published by Springer-Verlag Berlin Heidelberg New York in 1976
Softcover reprint of the hardcover 1st edition 1976

Typesetting: R. & J. Blank, München.

Contents

STRUCTURE AND BONDING is issued at irregular intervals, according to the material received. With the acceptance for publication of a manuscript, copyright of all countries is vested exclusively in the publisher. Only papers not previously published elsewhere should be submitted. Likewise, the author guarantees against subsequent publication elsewhere. The text should be as clear and concise as possible, the manuscript written on one side of the paper only. Illustrations should be limited to those actually necessary.

Manuscripts will be accepted by the editors:

Professor Dr. *Jack D. Dunitz*   Laboratorium für Organische Chemie der Eidgenössischen Hochschule
                                     CH-8006 Zürich, Universitätsstraße 6/8

Professor Dr. *Peter Hemmerich*   Universität Konstanz, Fachbereich Biologie
                                     D-7750 Konstanz, Postfach 733

Professor *James A. Ibers*   Department of Chemistry, Northwestern University
                                     Evanston, Illinois 60201/U.S.A.

Professor Dr. *C. Klixbüll Jørgensen*   51, Route de Frontenex,
                                     CH-1207 Genève

Professor *Joe B. Neilands*   University of California, Biochemistry Department
                                     Berkeley, California 94720/U.S.A.

Professor Dr. *Dirk Reinen*   Fachbereich Chemie der Universität Marburg
                                     D-3550 Marburg, Gutenbergstraße 18

Professor *Robert Joseph P. Williams*   Wadham College, Inorganic Chemistry Laboratory
                                     Oxford OX1 3QR/Great Britain

SPRINGER-VERLAG   SPRINGER-VERLAG
                                                         NEW YORK INC.

D-6900 Heidelberg 1   D-1000 Berlin 33
P. O. Box 105280   Heidelberger Platz 3   175, Fifth Avenue
Telephone (06221) 487·1   Telephone (030) 822001   New York, N. Y. 10010
Telex 04-61723   Telex 01-83319   Telephone 673-2660

Paradoxical Violations of Koopmans' Theorem, with Special Reference to the $3d$ Transition Elements and the Lanthanides

Ricardo Ferreira

Departamento de Fisica, Universidade Federal de Pernambuco, Cidade Universitária, 50.000 Recife-Pe, Brasil

Table of Contents

I. Introduction

The tendency of chemists to rationalize chemical phenomena in terms of atomic parameters is an inheritance from the dominating influence of Analytical Chemistry, in the Lavoisierian sense of breaking-down substances into their constituent elements, in the course of the history of Chemistry. The success of the chemical atomic theory itself resulted, in a fundamental sense, from the manifestly discontinuous character of gross chemical changes (1). Subtler chemical processes and spectroscopic phenomena reveal that chemistry is, to a large extent, a question of the influence of the adjacent atoms (2). The mutual influence of adjacent atoms and the interaction of these atoms with electromagnetic fields, treated in the semi-classical approach, can be described by quantum-mechanical perturbation theory, and *Dewar* has commented (3) that "one could regard Chemistry in general as an exercise in perturbation theory on the part of Nature". Relativistic quantum mechanics in the sense of Dirac's formalism becomes important for high Z values (4), and these effects become chemically significant for the lanthanides. Although chemists are modest scientists, and it is argued that valence and optical transition electrons (1—20 eV) are far from having relativistic velocities, the detailed interpretation of some processes seems to require methods of the many-body perturbation theory (5). These methods represent a radically different view of processes such as excitations and ionizations from the usual "orbital and electrons" description. It is quite possible that we are approaching a situation in which the really meaningful questions depend on second quantitization concepts and methods. This certainly makes chemists uneasy because it is a social phenomenon that *chemical research* will close down when the remaining problems cannot be raised by chemical concepts nor solved by chemical techniques. This paper can be seen as a tentative to preserve the concept of orbital energy as applied to $3d$ transition elements and lanthanides.

There is little doubt that chemists readily accepted the Valence Bond description of molecules because VB wave-functions are electron-pair wave-functions and therefore easy to translate mentally into the electron-pairing scheme of *Lewis*. *Dewar* (6) has quoted *Longuet-Higgins* as saying that chemists did not gain much from the resonance theory, but historically it furnished chemists with a bridge from classical structural theory (7). The quasi-transferability of electron-pair wave-functions (geminals) represents the most sophisticated support for the chemists' claims that molecules are made of atoms (8). Thus, the lone and bonding pairs of the oxygen atoms are preserved in OH, H_2O, and H_2O_2, at least to a remarkable degree (9). Molecular orbitals, delocalized around the nuclei, are not expected to have this transferability property. The classical example is the *James* and *Coolidge* wave-function for the H_2 molecule (10) which shows no relation whatsoever with the hydrogen atom eigenfunctions. Even the LCAO approach preserves only roughly the atomic orbitals. For example, the $2a_1$ orbital of the water molecule is very much unique, although the $1a_1$ orbital, being almost the $1s$ orbital of the oxygen atom, can be said to be present in other molecular species containing an oxygen atom. Chemists in general tended to regret this "going beyond the atoms themselves", to paraphrase *Kekulé (11)*, but as Chemistry became increasingly a question of excited states — in spectroscopy, photochemistry, chemical kinetics, and so on — the huge difficulties confronting VB theory forced chemists into the MO field. Rare-earth chemists, particularly rare-earth spectroscopists, are the least prone to use molecular orbital concepts. Their systems are essentially "free ion" systems to which one adds a Madelung potential, and since the spin-orbit splitting may encompass two or more atomic terms, they remain impregnable to MO arguments.

Rebuffed in their expectations of transferability of atomic parameters, chemists were glad to hang-on to fixed sequences of orbital energy levels, characteristic of each molecular species. The new hope, illustrated by the enormous popularity of orbital energy level diagrams (12), was supported by the expectation that these energy levels do not cross one another with changes in the orbital occupation numbers. This was a crucial point because all sorts of processes, such as electronic excitations, ionizations, reactive collisions, etc., can be interpreted in terms of changes in the orbital occupation numbers. The whole scheme depends on the validity of single determinantal representations as well as of Koopmans' theorem (13). This theorem is the statement that one-electron energy eigenvalues are exactly one-electron binding energies (orbital energies), the question of the identification of SCF eigenvalues with orbital energies having been discussed in many places (14).

Koopmans' statement is only approximately valid because it assumes that the SCF eigenfunctions are *frozen* with respect to changes in the occupation number of one or more of such eigenfunctions. In fact, electronic eigenfunctions relax with a half-life of the same order as that of the ionization process itself. The relaxation lowers the energy of the $(n-1)$-electron system, and, as pointed out by *Lorquet* (15), this "thawing energy" (16) makes ionization energies calculated by Koopmans' statement, $I(K) = -\epsilon^{SCF}$, larger than the *vertical* ionization energies. In molecules there also occurs rotational-vibrational relaxation and as a result *adiabatic* ionization energies are even lower than the corresponding vertical values, but orbital relaxation occurs in atoms as well, where in a sense there is only vertical ionizations (17). Ionization energies calculated from the difference in the total SCF electronic energy of the ion and the neutral species are represented by $I(\Delta_{SCF})$, and since both ionization energies and thawing energies, Δ_T, are positive quantities, we can write:

$$I(\Delta_{SCF}) = I(K) - \Delta_T < I(K) \qquad (1)$$

SCF eigenvalues include spin-correlation (the *Fermi* hole) but they differ from the correct non-relativistic energy by the electrostatic *correlation energy* (the *Coulomb* hole) (18). It is sometimes assumed (19, 20) that the correlation energy of a neutral atom or molecule, E^0_{corr}, defined as a positive quantity, is always larger than that of the corresponding positive ion, E^+_{corr}, but *Bagus* (17) have shown that in some cases (the $2s$ orbital of the Ne atom and the $3s$ orbital of the Ar atom, for example) calculated $I(\Delta_{SCF})$ values are larger than the corresponding experimental

3

ionization energies. Since, provided the spin correlation is taken correctly in both the closed- and open-shell cases, we can write:

$$I_{exp} = I(\Delta_{SCF}) + E^0_{corr} - E^+_{corr} \tag{2}$$

a calculated $I(\Delta_{SCF})$ larger than the experimental value indicates that $\Delta E_{corr} = E^0_{corr} - E^+_{corr} < 0$. With respect to Koopmans ionization energies, a positive ΔE_{corr} value is opposite to the thawing energy, since it means that correlation stabilizes the neutral system more than it does the positive ion. A negative ΔE_{corr}, on the other hand, will add to Δ_T in securing that $I(K)$ is larger than the vertical ionization energy. Experience indicates that in absolute values the thawing energy is larger than ΔE_{corr}, hence it is possible to write:

$$I_{exp} = I(K) + \Delta E_{corr} - \Delta_T < I(K) \tag{3}$$

Both *Lorquet* (*15*) and *Bagus* (*17*) have pointed out that although inequality (*1*) is quite general, the same cannot be said of expression (*3*), although it is supported by the available evidence.

Pedestrian violations of Koopmans' theorem in the sense of *Lorquet* was accepted with grace, the annoying possibilities of this fact only becoming audible when systems were found for which Koopmans' statement gives an incorrect description of the *ordering* of energy levels with respect to ionization energies. Good *ab initio* calculations are now available to prove that only many-body perturbation theory, such as second quantization methods, can predict electronic binding energies accurately enough. Consider, for example, the case of the nitrogen molecule, N_2 ($^1\Sigma_g^+$). Table 1 shows in the first column the experimental values of the three lowest vertical ionization energies (*21*). Koopmans' values, calculated from the *ab initio* eigen-values of *Cade et al.* (*22*) are shown in column 2. *Cade et al.* have also calculated the ionization energies by the Δ_{SCF} method, and these are shown in the third column. Finally, column 4 shows the SCF−$X\alpha$ values calculated by *Connolly* (*23*).

It is seen that no SCF method reproduces the experimental values correctly, although the SCF−$X\alpha$ calculation gives the correct ordering of the orbital energies.

Table 1. Ionization Energies of the Nitrogen (N_2) Molecule (in eV)

Experimental[a]	Koopmans[b]	Δ_{SCF} values[c]	SCF-X_α[d]
15.6 ($^2\Sigma_g$)	17.36 (3 σ_g)	16.01 (3 σ_g)	14.1 (3 σ_g)
16.9 ($^2\Pi_u$)	17.10 (1 π_u)	15.67 (1 π_u)	18.2 (1 π_u)
18.8 ($^2\Sigma_u$)	20.92 (2 σ_u)	19.93 (2 σ_u)	18.3 (2 σ_u)
[a]) Ref. (*21*)	[b]) Ref. (*22*)	[c]) Ref. (*22*)	[d]) Ref. (*23*)

On the other hand, *Cederbaum et al.* (*24*) have applied a second quantization method, the Green's function technique, to the N_2 problem, with excellent results. Unfortunately, second quantization methods are alien to the one-electron orbital concept; for example, it is clear that Cederbaum's successive approximations do not correspond to ever larger chunks of correlation energy being taken into account.

Starting with *Basch, Hollister* and *Moscowitz's* calculations of CuF_2 (*25*) there is an increasing number of reliable *ab initio* studies of transition metal compounds. We shall return to some of these results in *Section* III of the present paper, but we would like to stress that it was found in a certain number of cases that the one-to-one correspondence between Koopmans' ionization energies and the experimental vertical ionization energies breaks down. *Veillard et al.* (*26*) and *Cederbaum et al.* (*24*) call this failure of Koopmans' statement to predict the correct ordering of ionization energies simply a "breakdown of Koopmans' theorem". Considering, however, the pedestrian kind of breakdown, and following Professor Jørgensen's suggestion, I will call it a *paradoxical violation of Koopmans' theorem*. The first reported case of such violation of Koopmans' theorem was that of *Evans et al.* on the photo-electron spectra of $Mn(CO)_5 X(X = H, Cl, Br, I, CF_3, COCF_3, CH_3)$ (*27*). A clear-cut example is provided by the work of *Coutière, Demuynck* and *Veillard* on ferrocene (*28*). In this careful *ab initio* calculation it was found that in neutral ferrocene molecules the e_{2g} orbital with an eigenvalue of -14.4 eV is deeper than the e_{1u} orbital ($\epsilon_{e_{1u}} = -11.7$ eV). From Koopmans' theorem it follows that $I_{e_{2g}}(K) = 14.4\,\text{eV} > I_{e_{1u}}(K) = 11.7$ eV. However the difference in total energy between the neutral molecule of ferrocene with configuration ($\ldots a_{2u}^2\, a_{1g}^2\, e_{2g}^4\, e_{1g}^4\, e_{1u}^4$) and the positive ion ($\ldots a_{2u}^2\, a_{1g}^2\, e_{2g}^3\, e_{1g}^4\, e_{1u}^4$) is 8.3 eV, whereas the difference between the neutral molecule and the ion ($\ldots a_{2u}^2\, a_{1g}^2\, e_{2g}^4\, e_{1g}^4\, e_{1u}^3$) is 11.1 eV. These Δ_{SCF} values fit much better the experimental data of *Evans et al.* (*29*) than Koopmans' values. The improvement achieved with the Δ_{SCF} method over the *Koopmans'* ionization energies was ascribed (*28*) to differences in the thawing energies accompaning the removal of a metal-localized e_{2g} electron ($\Delta_T = 6.1$ eV) and a ligand-localized e_{1u} electron ($\Delta_T = 0.6$ eV), but it should be remarked that the Xα method, which does not include orbital relaxation, gives results that are better than the Δ_{SCF} values (*30*) explicitly.

To have found that there are severe limitations to the concept of a fixed orbital energy sequence for each system is disturbing to many chemists. As pointed out by *Hand, Hunt,* and *Schaefer*, who have found a similar breakdown of Koopmans' theorem in their *ab initio* calculations on D_{3h} FeF_3 (*31*), "it would be difficult for the theoreticians to convice the practising inorganic chemist to abandon his orbital energy level diagrams". It can be argued, for example, that the Hartree-Fock method is handicapped by the fact that in its standard form it is based on the minimization of the energy of a single determinantal wave-function; or that it is naive not to include configuration interaction, thereby mixing-up eigenfunctions and taking into account some correlation energy; and so on *ad nauseam*. The fact remains that the paradoxical violations of Koopmans' theorem is unwelcomed for practical reasons. The limitations of the concept of a fixed orbital sequence in molecules are of the

5

same type as those shown by the idea of a unique sequence of *atomic* orbital eigenvalues to be used in the Aufbau principle of atoms, but the practical consequences are potentially worse. The atomic orbital energy sequence is a function of Z and of the oxidation states of the elements, but there are only about 280 chemically meaningful oxidation states for all elements (*32*), whereas we know more than $2 \cdot 10^6$ molecular species and the number is steadily increasing. Since the factors that favour the paradoxical violation of Koopmans' theorem have not been diagnosed, it is difficult to predict in which cases it will occur. It has been suggested (*28, 32*) that the paradoxical situation is favoured by large Madelung potentials, small metal d of f orbitals with large interelectronic repulsion, and large ligand orbitals with low interelectronic repulsion term. Thus, the breakdown occurs in square planar $Ni(CN)_4^{2-}$ but not in tetrahedral $Ni(CO)_4$ (*33*), and it does occur in *tris*-hexafluoroacetylacetonate complexes (*34*), in ferrocene (*28*), and in rare-earth fluorides and oxides (*35, 36*). This situation affects all aspects of Chemistry. For example, the *frontier orbital* theory of reaction mechanisms (*37*) stresses the symmetry properties of the highest occupied MO (HOMO) and of the lowest occupied MO (LUMO) of the reacting species in the point group of the transition state. Suppose we are discussing the protonation of ferrocene; the question arises, what is the HOMO of the ferrocene molecule? Is it the e_{1u} orbital given by Koopmans' sequence of eigenvalues, or is it the e_{2g} orbital whose energy on protonation may follow the ionization pattern and be higher than that of the protonated e_{1u} orbital?

II. Concerning One-Electron Eigenvalues

We have recalled that *Lorquet* (*15*) firstly pointed out that ionization energies obtained from Koopmans' theorem are always *larger* than calculated $I(\Delta_{SCF})$ values and experimental ionization energies (Eq. (1) and (3)). Let ϵ_μ^{SCF} and ϵ_ν^{SCF} be two energy eivenvalues of a system, and we will suppose that $|\epsilon_\mu^{SCF}| > |\epsilon_\nu^{SCF}|$. If we define the Koopmans' defects $\delta_\mu = -\epsilon_\mu^{SCF} - I_\mu(\exp)$ and $\delta_\nu = -\epsilon_\nu^{SCF} - I_\nu(\exp)$ the condition:

$$\epsilon_\nu^{SCF} - \epsilon_\mu^{SCF} < \delta_\mu - \delta_\nu = \epsilon_\nu^{SCF} - \epsilon_\mu^{SCF} + I_\nu(\exp) - I_\mu(\exp) \qquad (4)$$

characterizes a situation in which $I_\nu(\exp) > I_\mu(\exp)$ but $I_\mu(K) > I_\nu(K)$. We say that there occurs a paradoxical violation of *Koopmans'* theorem (*38*). Examples of such behavior have accumulated during the last few years. Thus, it was found that the ionization energy of a half-filled $3d$ orbital in tris-hexafluoroacetylacetonate of Fe(III) is larger than the ionization energy of the loosest bonded ligand orbital (*34*). It was also found (*35, 36*) that the ionization energies of the half-filled $4f$ orbitals of some rare-earth fluorides and oxides are larger than those of the ligand orbitals. In the same way, in rare-earth antimonides, MeSb, the antimony $5p$ ionization potentials are smaller than those of the $4f$ electrons (*39*). In terms of the accepted orbital energy diagrams for such compounds these facts mean that partly filled anti-bonding orbitals may have higher ionization energies than their bonding counterparts (*40*). This paradox has been collo-quially described as "the third revolution in ligand field theory" (*41, 42*). Other para-doxical situations occur in metallic lanthanides, where the ionization energies of the half-filled $4f$ orbitals are larger than those of the *Fermi* levels (*40, 43*). A similar situa-tion exists in gaseous FeF_3, according to the *ab initio* calculations of *Hand et al.* (*31*): the half-filled $1e''$, $6a_1'$ and $5e'$ orbitals lie below the fully occupied $7a_1'$, $6e'$, $7e'$, $2e''$, $3a_2''$, and $4a_2'$ orbitals. These results lead the authors to emphasize the limitations of the orbital energy levels diagrams: how can we explain that the configuration $(\dots (1e'')^2$ $(6a_1')^1 (5e')^2 (7a_1')^2 (6e')^4 \dots)$ is lower than, for example, $(\dots (1e'')^2 (6a_1')^2$ $(5e')^2 (7a_1')^1 (6e')^4 \dots)$? Likewise, what prevents the conduction electrons of metallic lanthanides from invading the $4f$ shell?

The full implications of the fact that one-electron SCF eigenvalues are, in semi-empirical schemes independent-particle eigenvalues are not always realized. In SCF–HF methods the interelectronic repulsion is taken in an average way, and from this feature results that electrons of a closed-shell move as independent particles, although this is concealed by the peculiarity that SCF–HF methods do not take orbital occupa-tion numbers explicitly into consideration. SCF energy eigenvalues are, except for the correlation energy, the correct non-relativistic one-electron energies, but application of Koopmans' theorem will give ionization energies which, apart from orbital relaxa-tion, differ by $\frac{1}{2} J_{\mu\mu}$ from the correct one-electron binding energies, $J_{\mu\mu}$ being the electron repulsion integral. This often-neglected fact was recognized by *Dewar, Hash-*

mall and *Venier* in their original paper on the "half-electron method" (*47*); the method was put on a firm foundation by *Kollmar* (*48*), and more recently *Dewar, Kollmar* and *Suck* (*49*) have made increasing use of it.

The expression of the electronic SCF energy of a closed-shell system is:

$$E = 2 \sum_{i=i}^{N} E_i^c + \sum_i^N \sum_j^N (2 J_{ij} - K_{ij}) \tag{5}$$

The contribution to expression (5) from the kth electron is:

$$E_k = E_k^c + \sum_{i=i}^{N} (2 J_{ki} - K_{ki}) \tag{6}$$

Supposing that the orbitals remain frozen the energy of the resulting positive ion is $E^+ = E - E_k$, and, therefore, $E^+ - E = I(K) = -E_k$. This is, of course, a statement of Koopmans' theorem. It is not always recognized that although expression (6) represents the contribution of the kth *electron* to the energy E, the summation in (5) is over the N *orbitals*. This means that the two electrons in each doubly-occupied orbital contribute equally to the energy E, and, if Koopmans' theorem is applied, have the same binding energy. In other words, the interelectronic repulsion in a given orbital is taken as spanning uniformly in the interval $2 \geqslant n \geqslant 0$, although the ionization of a closed-shell corresponds rather to a transition from $n = 2$ to $n = 1$. Clearly, and quite apart from orbital thawing, ionization energies calculated from Koopmans' statement will be *too large* by $\frac{1}{2} J_{\mu\mu}$. This can be proved if we write the expression for the SCF electronic energy of a closed-shell system as a function of the orbital occupation numbers. As shown by *Dewar et al.* (*47*) Eq. (5) transforms to:

$$E = \sum_\mu n_\mu E_\mu^c + \frac{1}{4} \sum_\mu \sum_\mu n_\mu n_\nu (2 J_{\mu\nu} - K_{\mu\nu}) \tag{7}$$

For closed-shell systems ($n_\mu = n_\nu = 2$) Eq. (7) reduces to (5). However, the contribution to (7) of one electron in orbital μ is given by:

$$E_\mu = E_\mu^c + \frac{1}{2} \sum_\mu (2 J_{\mu\nu} - K_{\mu\nu}) = E_\mu^c + \frac{1}{2} J_{\mu\mu} \tag{8}$$

By Koopmans' theorem only half the interelectronic repulsion is computed in the calculation of the ionization energy, which will be therefore too large by $\frac{1}{2} J_{\mu\mu}$. *Dewar, Hashmall* and *Venier* (*47*) have also shown that if one treats the closed-shell and the open-shell systems separately, but if the open-shell system is also described by s single Slater determinant, an error of $\frac{1}{4} J_{\mu\mu}$ remains in the calculation of binding energies.

Thus, if ψ_0 is the single-occupied orbital, and ψ_μ, ψ_ν, etc., are doubly-occupied ones, the SCF electronic energy (Eq. (5)) can be written:

$$E^+ = 2\sum_\mu E_\mu^c + E_0^c + \sum_\mu \sum_\nu (2J_{\mu\nu} - K_{\mu\nu}) + \sum_\mu (2J_{\mu 0} - K_{\mu 0}) \tag{9}$$

If one writes (9) as a function of the orbital occupation numbers it is easy to show that:

$$E^+ = \sum_\mu 2E_\mu^c + E_0^c + \tfrac{1}{4}\{\sum_\mu \sum_\nu 4(2J_{\mu\nu} - K_{\mu\nu}) + \sum_\mu 2(2J_{\mu 0} - K_{\mu 0}) +$$

$$+ \sum_\nu 2(2J_{\nu 0} - K_{\nu 0}) + (2J_{00} - K_{00})\} = 2\sum_\mu E_\mu^c + E_0^c + \tag{10}$$

$$+ \sum_\mu \sum_\nu (2J_{\mu\nu} - K_{\mu\nu}) + (2J_{\mu 0} - K_{\mu 0}) + \tfrac{1}{4}J_{00}$$

E^+ according to (10) is larger than E^+ according to (9) by a term $\tfrac{1}{4}J_{00}$. For the neutral, closed-shell system, on the other hand, Eq. (7) and (5) give the same value of E. Therefore the difference $E^+ - E$, that is, the ionization energy, will be too large by the amount $\tfrac{1}{4}J_{00}$. No such error appears if the open-shell system is described by a sum of *Slater* determinants (*50*).

A great deal of understanding about interelectronic repulsion energies is gained if one uses an approach explicitly showing the dependence of orbital energies with occupation numbers. This is the approach considered by the SCF–X_α and related methods (*51–54*) and to these developments we turn now our attention.

In the one-electron model it is reasonable to disregard multiplet separations and assign all the multiplet levels to the average energy of their barycenters. The expression giving the average energy of all multiplets arising from a given configuration of an atom (*51*) is:

$$E = \sum_\mu \{-n_\mu I_\mu + \tfrac{1}{2}n_\mu(n_\mu - 1)J_{\mu\mu}\} + \sum_{\mu > \nu} \sum_\nu n_\mu n_\nu (\mu, \nu) \tag{11}$$

The indices μ and ν indicate atomic orbitals, and the quantities I_μ, $J_{\mu\mu}$ and (μ, ν) are, respectively, the sum of the kinetic and the potential energies in the field of the nucleus, the electrostatic interaction between electrons in orbital ψ_μ, and the Coulomb and exchange interactions between electrons in orbitals μ and ν, averaged over all possible pairs of quantum numbers m_l and m_s of the two electrons (*51*). It is easily shown (*51, 52*) that the binding energy of a doubly occupied orbital, given by the

9

difference $\{E(n_\mu) - E(n_\mu - 1)\}_{n_\mu = 2}$ is exactly equal to $\left(\dfrac{\delta E}{\delta n_\mu}\right)_{n_\mu = 1.5}$, and that the binding energy of an open-shell system, given by $\{E(n_\mu) - E(n_\mu - 1)\}_{n_\mu = 1}$ is exactly equal to $\left(\dfrac{\delta E}{\delta n_\mu}\right)_{n_\mu = 0.5}$. The calculation of orbital ionization energies as the derivatives of the total energy E with respect to the occupation numbers at the values $n_\mu = \frac{3}{2}$ and $n_\mu = \frac{1}{2}$ forms the bases of the *transition state* technique of the X_α methods. These methods, then, presuppose a continuous function of E with respect to n. It is not always recognized that the one-determinant SCF energy expression showing explicitly the dependence on occupation numbers, Eq. (7), does not correspond to Eq. (11), but to the expression:

$$E = \sum_\mu \{ - n_\mu I_\mu + n_\mu^2 / 4 J_{\mu\mu} \} + \sum_{\mu > \nu} \sum_\mu n_\mu n_\nu (\mu, \nu) \tag{12}$$

It is interesting that although this fact is known for more than a decade (*55–58*) its obvious connections with the "half-electron" method of *Dewar* (*47*) have never been pointed out before. At about the same time that *Dewar* published his first paper on the "half-electron" method (*59*) we did show (*60*) that Eq. (12) corresponds to treat the interelectronic repulsion as the interaction between two half-electron-pairs, *which is not zero for one electron* (two half-electrons). For closed-shell systems Eq. (11) and (12) give the same value for E, but for $n_\mu = 1$ the value of E according to (12) is not the correct value

$$- I_\mu + \sum_\nu n_\nu (\mu, \nu),$$

such as given by Eq. (11), but

$$- I_\mu + \frac{1}{4} J_{\mu\mu} + \sum_\nu n_\nu (\mu, \nu).$$

The ionization energy calculated from Eq. (12) is therefore $\frac{1}{4} J_{\mu\mu}$ larger than that calculated from Eq. (11). This can also be seen from the fact that the derivative of E with respect to n_μ according to (12) is

$$\left(\frac{\delta E}{\delta n_\mu}\right) = - I_\mu + \frac{1}{2} n_\mu J_{\mu\mu} + \sum_{\nu \neq \mu} n_\nu (\mu, \nu) \tag{13}$$

and that its value for $n_\mu = \frac{3}{2}$ is larger than the value given by the derivative of (11) exactly by $\frac{1}{4} J_{\mu\mu}$.

Eq. (11) for $n_\mu = 1$ gives the correct energies (*61*) whereas Eq. (12) furnishes values which are too high by $\frac{1}{4} J_{\mu\mu}$. Application of Koopmans' theorem, however, whether to Eq. (11) or (12), will give ionization energies that are too large by $\frac{1}{2} J_{\mu\mu}$. This is simply because the energy of a doubly occupied orbital is, according to both equations,

$$- 2I_\mu + J_{\mu\mu} + 2 \sum_{\nu \neq \mu} n_\nu(\nu, \mu) ,$$

the one-electron eigenvalue being half that value:

$$- \epsilon_\mu = I_\mu(K) = -\frac{E}{2} = I_\mu - \frac{1}{2} J_{\mu\mu} - \sum_{\nu \neq \mu} n_\nu(\nu, \mu). \tag{14}$$

In Eq. (14) the nuclear interaction term (I_μ) and the inter-orbital electron repulsion terms $(\sum_{\nu \neq \mu} n_\nu(\nu, \mu))$ are correct, but it includes just half of the intra-orbital repulsion terms. In Fig. 1 we show graphically the behavior of the orbital energy ϵ_μ according to (11) (curve *c*) and to (12) (curve *b*). Curve *a* shows the behavior of the independent-particle orbital.

The independent particle function (curve *a*) is secant to both curves *b* and *c* and it has a constant slope of $-I_\mu + \sum_{\mu \neq \nu} n_\nu(\mu, \nu) + \frac{1}{2} J_{\mu\mu}$, which is expression (14) with

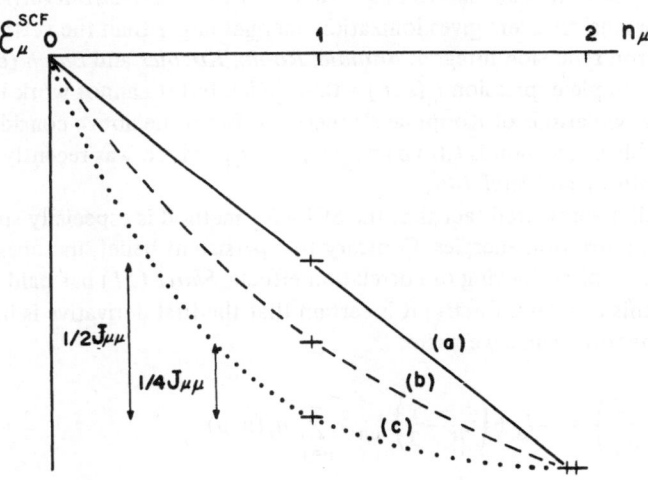

Fig. 1. Orbital Energies and Occupation Numbers.

Curve *a*: $\epsilon_\mu = -n_\mu I_\mu + \sum_\nu \sum_\mu n_\mu n_\nu(\mu, \nu) + \frac{1}{2} n_\mu J_{\mu\mu}$

Curve *b*: Eq. (12) of the text. Curve *c*: Eq. (11) of the text.

the opposite sign. This is the same value of the derivatives of Eq. (11) and (12), curves c and b, at the point $n_\mu = 1$. If ψ_μ is an *atomic orbital*, the derivative

$$\left(\frac{\delta E_\mu}{\delta n_\mu}\right)_{n_\mu=1}$$

according to both (11) and (12), is clearly *Mulliken's* electronegativity of orbital ψ_μ (62–65). Using Moffitt's approximation for the interelectronic repulsion integral:

$$J_{\mu\mu} = I_\mu - \Sigma_{\nu \neq \mu}\, n_\nu(\nu, \mu) - A_\mu \tag{15}$$

where A_μ is the electron affinity of orbital ψ_μ, we can write:

$$\begin{aligned}
\left(\frac{\delta E_\mu}{\delta n_\mu}\right)_{n_\mu=1} &= -I_\mu + \sum_{\nu \neq \mu} n_\mu(\nu, \mu) + \tfrac{1}{2} J_{\mu\mu} = \\
&= -I_\mu + \sum_{\nu \neq \mu}(\mu, \nu) + \tfrac{1}{2}\{I_\mu - \sum_{\nu \neq \mu} n_\nu(\nu, \mu) - A_\mu\} = \\
&= -\tfrac{1}{2}\{I_\mu - \sum_{\nu} n_\nu(\nu, \mu) + A_\mu\}
\end{aligned} \tag{16}$$

We come to the conclusion that the one-electron eigenvalue for an independent-particle model is equal to the electronegativity of the orbital, and although Eq. (11) and (12) refer to atomic orbitals there is no reason why we cannot generalize this statement. Because the one-electron eigenvalues are *orbital electronegativities*, application of Koopmans' theorem gives ionization energies larger than the vertical ones by half the electron repulsion integral. *Brundle, Robin, Kuebler* and *Basch (66)* have proposed the simple expression $I_\mu(exp) = 0.92\, I_\mu(K)$, but it cannot work in the cases of paradoxical violations of Koopmans's theorem. From the above considerations a more reasonable expression is $I_\mu(exp) = I_\mu(K) - \tfrac{1}{2} J_{\mu\mu}$, which was recently proposed by *Dewar, Kollmar* and *Suck (49)*.

It is a well documented fact that the SCF–X_α method is especially suited for the calculation of ionization energies. Contrary to a persistent belief, its sucess is not due to inclusion of orbital thawing or correlation effects. *Slater (51)* has paid particular attention to this problem. Firstly, it is certain that the first derivative is independent of orbital relaxation, since we write:

$$\left(\frac{\delta E}{\delta n_\mu}\right) = -I_\mu + \left(n_\mu - \tfrac{1}{2}\right) J_{\mu\mu} + \sum_{\nu \neq \mu} n_\nu(\nu, \mu) \tag{17}$$

for the derivative of E with respect to n_μ, and since E is given by the summation (11), this implies that the derivatives

$$\left(\frac{\delta I_\nu}{\delta n_\mu}\right), \quad \left(\frac{\delta I_\eta}{\delta n_\mu}\right), \quad \text{etc.,}$$

as well as

$$\left(\frac{\delta J_{\nu\nu}}{\delta n_\mu}\right), \quad \left(\frac{\delta J_{\eta\eta}}{\delta n_\mu}\right), \quad \text{etc.,}$$

are all zero. To quote *Slater (51)*: "The change in total energy when the occupation numbers are changed arises from two causes, the explicit dependence of energy on occupation numbers, and the change in energy on account of modification of the orbitals with occupation number. To the first order of small quantities, the second effect vanishes on account of the use of the variational method, and this leaves the same change in total energy, to the first order, whether or not relaxation is considered. "This is true for the X_α method as well as for the *Hyper-Hartree-Fock* (HHF), or *Hartree-Fock-Slater* (HFS) method. In this technique the ionization energy is given by a power series expansion:

$$\{E(n_\mu) - E(n_\mu - 1)\} = \left(\frac{\delta E}{\delta n_\mu}\right) - \frac{1}{2}\left(\frac{\delta^2 E}{\delta n_\mu^2}\right) + \frac{1}{6}\left(\frac{\delta^3 E}{\delta n_\mu^3}\right) - \dots \quad (18)$$

It is easily shown that:

$$\left(\frac{\delta E}{\delta n_\mu}\right)_{n_\mu = 3/2} = \left(\frac{\delta E}{\delta n_\mu}\right)_{n_\mu = 2} - \frac{1}{2}\left(\frac{\delta^2 E}{\delta n_\mu^2}\right)_{n_\mu = 2} \quad (19)$$

and

$$\left(\frac{\delta E}{\delta n_\mu}\right)_{n_\mu = 1/2} = \left(\frac{\delta E}{\delta n_\mu}\right)_{n_\mu = 1} - \frac{1}{2}\left(\frac{\delta^2 E}{\delta n_\mu^2}\right)_{n_\mu = 1} \quad (20)$$

which means that the transition-state concept of the X_α method corresponds to taking the second order term of the expansion in the HFS method. Orbital relaxation cannot therefore be evoked to explain the success of those new methods.

As to the correlation energy, all integrals in Eq. (11) are over non-correlated orbitals. It is true that by taking the derivative of the average of a multiplet system we are implicitly by-passing the limitation of the spin-correlated *Hartree-Fock* single determinantal formalism, but with so many approximations involved in the SCF–X_α method it is not easy to judge how important is this particular feature.

It is difficult to avoid the conclusion that the success of the X_α and related methods in calculating ionization energies is mainly due to the correct expression for the interelectronic repulsion term $\frac{1}{2} n_\mu(n_\mu - 1) J_{\mu\mu}$, in contrast with the expression $\frac{1}{4} n_\mu^2 J_{\mu\mu}$ which occurs in SCF methods based on single determinantal molecular wave-functions. The transition state concept, instrumental in taking the correct interelectronic repulsion term, is therefore central to the success of the SCF–X_α method.

III. Independent-Particle Orbitals and Energy Level Diagrams

Chemists have learned much from independent-particle molecular orbital theories, such as *Hückel's*. The usefulness of orbital energy level diagrams (*12*) cannot be over-emphasized, but their limitations should be pointed out (*31*). These diagrams are based on what is essentially an independent-particle formalism, that is, on the premise that ϵ_μ/n_μ remains constant in the interval $2 \geqslant n_\mu \geqslant 0$. In fact, we may define an "independent-particle orbital" as an orbital for which the orbital energy $\epsilon_\mu(n_\mu)$ is linearly dependent on n_μ in that interval. For such orbitals the derivative $\left(\dfrac{\delta\epsilon_\mu}{\delta n}\right)$ is a constant equal to $-I_\mu + \sum_{\nu \neq \mu} n_\nu(\nu,\mu) + \frac{1}{2} J_{\mu\mu}$ (curve *a* of Fig. 1). The ionization energy calculated from Koopmans' statement, $I_\mu(K)$, is equal to $\left(\dfrac{\delta\epsilon_\mu}{\delta n_\mu}\right)$. It is well known (*55, 67*) that it is impossible for any system containing more than one electron to have its energy linearly dependent on the occupation number. This fact is a severe limitation of the independent-particle formalism, and we attribute the success of the SCF–X_α method mainly to its correct handling, through the transition state concept, of the n_μ^2 term. This entails the reasonable assumption that *differences* in the values of $(\Delta E_{corr} - \Delta_T)$ upon the removal of electrons from two differing orbitals are an order of magnitude smaller than the error in the independent-particle formalism, provided both orbitals are in the valence shell.

Consider now two orbitals, ψ_μ and ψ_ν, such that $|\epsilon_\mu^{SCF}| > |\epsilon_\nu^{SCF}|$. Fig. 2 shows the dependency of ϵ_μ^{SCF} and ϵ_ν^{SCF} on the occupation numbers. Curves a_μ and a_ν correspond

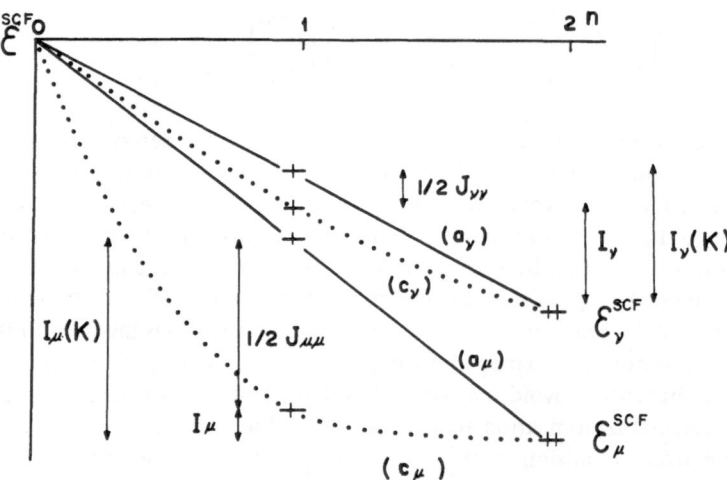

Fig. 2. Paradoxical breakdown of Koopmans' theorem. $I_\nu(K) < I_\mu(K)$, but $(IE)_\nu > (IE)_\mu$. Condition: $\{\epsilon_\nu^{SCF} - \epsilon_\mu^{SCF}\} < \frac{1}{2}\{J_{\mu\mu} - J_{\nu\nu}\}$

to the independent-particle approximation, with constant slopes equal to, respectively, $\epsilon_\mu(2)/2$ and $\epsilon_\nu(2)/2$. We have seen that from Koopmans' theorem $\epsilon_\mu(2)/2 = \epsilon_\mu^{SCF} = I_\mu(K)$, and $\epsilon_\nu(2)/2 = \epsilon_\mu^{SCF} = I_\nu(K)$, and since ϵ_μ^{SCF} is larger (more negative) than ϵ_ν^{SCF}, we will have $I_\mu(K) > I_\nu(K)$. Curves c_μ and c_ν represent the eigenvalues ϵ_μ^{SCF} and ϵ_ν^{SCF} according to Eq. (11). At the point $n_\mu = n_\nu = 1$, $a_\mu - c_\mu = \frac{1}{2} J_{\mu\mu}$, and $a_\nu - c_\mu = \frac{1}{2} J_{\nu\nu}$. The ionization energies according to Eq. (11) are:

$$(IE)_\mu = \{\epsilon_\mu^{SCF}(2) - \epsilon_\mu^{SCF}(1)\} = -\left(\frac{\delta\epsilon_\mu^{SCF}}{\delta n_\mu}\right)_{n_\mu=1.5} \tag{19}$$

$$(IE)_\nu = \{\epsilon_\nu^{SCF}(2) - \epsilon_\nu^{SCF}(1)\} = -\left(\frac{\delta\epsilon_\nu^{SCF}}{\delta n_\nu}\right)_{n_\nu=1.5} \tag{20}$$

In the case shown in Fig. 2, although $I_\mu(K) > I_\nu(K)$, we have $(IE)_\mu < (IE)_\nu$. This crossing-over of the one-electron eigenvalues and ionization energies will occur for any pair of eigenvalues for which the difference $\epsilon_\nu^{SCF} - \epsilon_\mu^{SCF}$ is *smaller* than $\frac{1}{2}\{J_{\mu\mu} - J_{\nu\nu}\}$. A pair of eigenvalues obeying this condition may well be a bonding orbital and its anti-bonding counter-part. It is this condition that determines the paradoxical breakdown of *Koopmans'* theorem *(24, 26)*. In chemical terms a paradoxical violations of Koopmans' theorem is likely to occur if the following conditions obtain:

(i)   small orbital energy differences;
(ii)   large interelectronic repulsion term of the lower (more negative) orbital;
(iii)   small interelectronic repulsion term of the higher (less negative) orbital.

A fourth condition, to be discussed later, is a large interatomic Coulomb potential *(Madelung* term) associated with the higher orbital. As kindly pointed out to me by *Dr. Alan Williams*, the first three conditions are also likely to occur only in polyatomic systems: in atoms, if two eigenvalues are close from one another (condition i), the interelectronic repulsion terms tend to be alike and therefore conditions (ii) and (iii) do not obtain.

It is clear from the diagram of Fig. 2 that if $\epsilon_\nu^{SCF} - \epsilon_\mu^{SCF} < \frac{1}{2}\{J_{\mu\mu} - J_{\nu\nu}\}$ the electron configuration $(\psi_\mu)^1 (\psi_\nu)^2$ is more stable than the configuration $(\psi_\mu)^2 (\psi_\nu)^1$, although ϵ_μ^{SCF} is lower than ϵ_ν^{SCF}. Such situations exist, for example, in metallic lanthanides *(36, 43)* and in gaseous D_{3h} symmetry FeF_3 *(31)*. They are paradoxical only in terms of independent-particle orbital energy diagrams. In the same way, if we are guided by the usual orbital energy level diagrams for $3d$ transition metal complexes and its rough extension to lanthanide complexes, it is possible in some cases to have partly filled antibonding orbitals with higher ionization energies than their doubly occupied bonding counterparts *(40)*. These are the paradoxical situations that characterize the "third revolution" in ligand field theory *(36, 41)*. Again, in inspection of the diagram of Figure 2 shows that this situation requires only that

$$\epsilon_\nu^{SCF} - \epsilon_\mu^{SCF} < \tfrac{1}{2}\{J_{\mu\mu} + J_{\nu\nu}\} \; ,$$

a much less stringent condition than

$$\epsilon_\mu^{SCF} - \epsilon_\mu^{SCF} < \tfrac{1}{2}\{J_{\mu\mu} - J_{\nu\nu}\} \; .$$

However, the systems that illustrate the "third revolution" are those characterized by the same four conditions discussed above and determined by

$$\epsilon_\nu^{SCF} - \epsilon_\mu^{SCF} < \tfrac{1}{2}\{J_{\mu\mu} - J_{\nu\nu}\} \; .$$

We are therefore led to propose a modification in the usual MO energy level diagrams of transition metal complexes which will enable us to encompass the "third revolution" cases with the other paradoxical cases.

Note added in proofs:

If we take $\left(\dfrac{\partial E}{\partial n_\mu}\right)$ in Eq. (17) for $n_\mu = \dfrac{3}{2}$ the resulting one-electron eigen-value conforms to Koopmans' theorem, but the sum of such values does not add to the total energy of the system. If we choose the value $n_\mu = 1$ in Eq. (17) the resulting orbital electronegativities add up to the correct total energy of the system, but they differ from Koopmans' theorem values by $\tfrac{1}{2} J_{\mu\mu}$.

IV. Molecular Orbital Energy Level Diagrams and The Third Revolution in Ligand Field Theory

The gas phase photo-electron spectra of the *tris*-hexafluoroacetylacetonates of the first row transition elements show an increase in the ionization energies of the $3d$ electrons from Ti(III) to Co(III). In the Fe(III) complex the ionization energy of the half-filled $3d$ orbital is larger than that of the loosest bound doubly occupied ligand orbital (*34*). Fig. 3 shows the energy level diagram of the octahedral complex Fe(III)-*tris*-hexafluoroacetylacetonate, showing, in the standard way (*68*), the ligand σ-orbitals with a lower diagonal element than the $3d$ metal orbitals.

We have shown that such situations will occur whenever $E_\nu^{SCF} - \epsilon_\mu^{SCF} < \frac{1}{2}(J_{\mu\mu} + J_{\nu\nu})$, and since ψ_μ is the bonding e_g orbital and ψ_ν is the anti-bonding e_g^* orbital, the situation could be described as paradoxical. However, the conditions leading to this type of paradox are quite non-specific and, in particular, do not apply in this case nor in those of the rare-earth fluorides and oxides (*69*). A single pattern of paradoxical behavior emerges, on the other hand, if the metal $3d$ orbital diagonal element is considered to be lower (more negative) than that of the outermost ligand σ-orbitals.

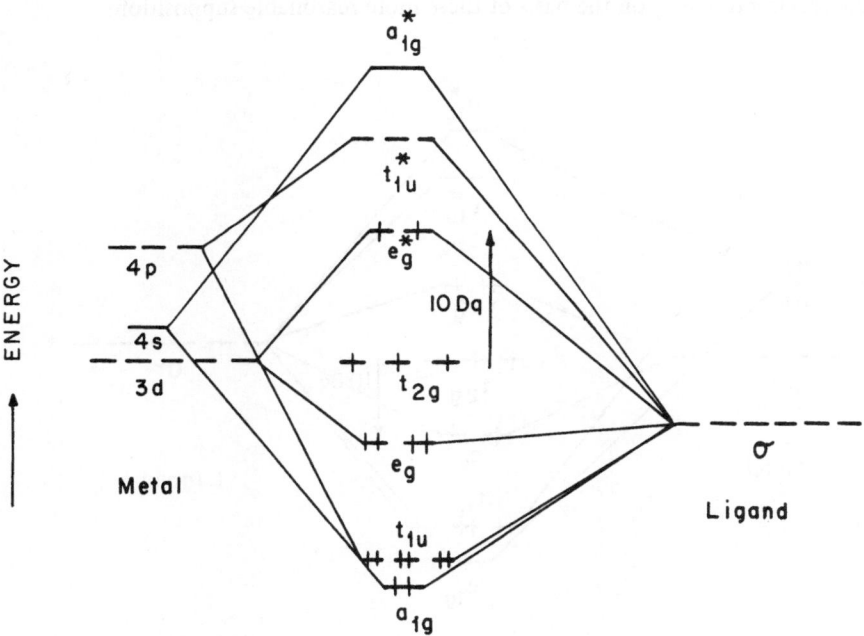

Fig. 3. Simple MO energy level diagram of the outermost σ-electron system of Fe(III) (hexafluoroacac)₃.

The reasons to consider the ligand σ-orbitals lower than the transition-metal $3d$ orbitals seem to be twofold. In the first place metal ionization energies tend to be smaller than those of typical a-metallic ligands. The second reason is a psychological one: covalency effects were added by chemists to the crystal field model, and this, describing a metal complex as purely ionic, requires infinitely large negative diagonal elements for the ligand orbitals. The need for ligand diagonal elements more negative than the metal's was later extended, in what was agreed to be a natural way, to neutral ligands.

As for the first reason, for high oxidation states (typically the $+3$ state of transition metals and lanthanides) the diagonal element of the metal is closer to an average of the successive ionization energies and should be lower than, for example, the $2p_\sigma$ orbital of oxygen or fluorine. This assumption is supported by the population analysis results of recent *ab initio* calculations of metal complexes. Thus, *Hand, Hunt* and *Schaeffer (31)* have found that the $1e''$, $6a_1'$ and $5e'$ orbitals of FeF_3, predominantly iron $3d$ orbitals, lie lower than the $7a_1'$ orbital, which is predominantly a fluorine $2p_\sigma$ orbital. The $7a_1'$, $6e'$, $2e''$, $3a_2''$ and $1a_2'$ orbitals, less negative than the $1e''$, $6a_1'$ and $5e'$ orbitals, contain the iron $4s$ and $4p$ orbitals mixed with the fluorine $2p_\sigma$ orbitals. A population analysis of the *ab initio* molecular orbitals of CrO_4^{2-} and MnO_4^- described by *Hillier* and *Saunders (70)* show that the metal $3d$ and the oxygen $2p$ orbitals contribute about equally to the $1e$ and $5t_2$ molecular orbitals. Fig. 4 shows a simplified MO energy level diagram of the system of Fe(III) (hexafluoroacetylacetonate)$_3$ on the basis of these more reasonable suppositions.

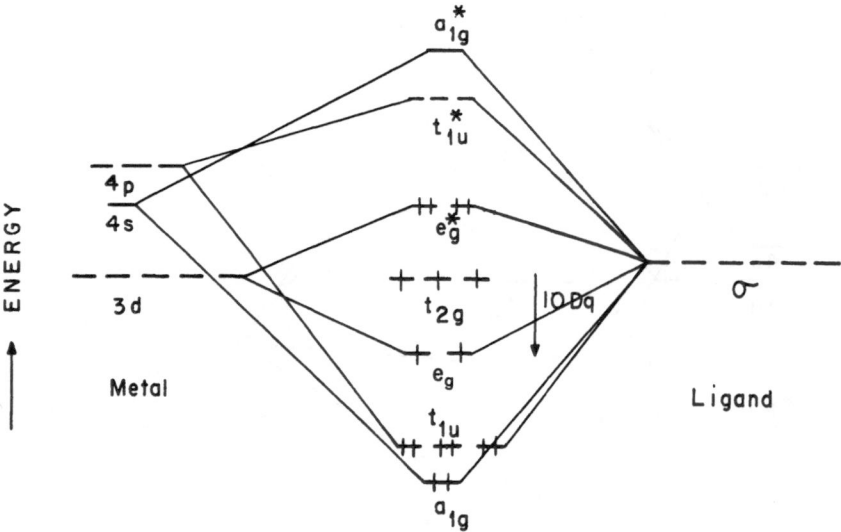

Fig. 4. MO energy leval diagram of the outermost σ-electron system of octahedral Fe(III) (hexafluoroacetylacetonate)$_3$, showing the ligand σ-orbital diagonal elements higher than the iron $3d$-orbital diagonal elements.

The paradoxical situation resides in the fact that the electronic configuration shown, $(\ldots (e_g)^2 \, (t_{2g})^3 \, (e_g{}^*)^4)$, is lower than $(\ldots (e_g)^4 \, (t_{2g})^3 \, (e_g{}^*)^2)$, although the e_g eigenvalue is more negative than the $e_g{}^*$ eigenvalue. We have shown that such situation obtains if:

$$I_{e_g}(K) - I_{e_g}{}^*(K) = \epsilon_{e_g{}^*}^{SCF} - \epsilon_{e_g}^{SCF} < \tfrac{1}{2} \{ J_{e_g} - J_{e_g}{}^* \} \tag{21}$$

This is in ample accord with the fact that the e_g orbitals with a considerable $3d$ character have a large interelectronic repulsion term (condition ii) and that this same term is small in the antibonding $e_g{}^*$ orbital, which is mainly a ligand orbital (condition iii). The ligand-to-metal charge transfer band is described in the new scheme as a $e_g{}^* \to e_g$ transition, not as a $e_g \to e_g{}^*$ transition, but, of course, it is the $e_g{}^*$ orbital that is now a predominantly ligand orbital.

The orbital energy level diagram of Figure 4 shows the ligand-field splitting parameter, $10\,Dq$, as a negative quantity. This just shows the limitations of such diagrams. Although $\epsilon_{e_g}^{SCF}$ is lower than $\epsilon_{t_{2g}}^{SCF}$, these orbitals are all half-filled, and the $10\,Dq$ parameter is a positive quantity if the second ionization energy of a t_{2g} orbital is larger than the first ionization energy of a e_g orbital. We have seen that this obtains if

$$\epsilon_{t_{2g}}^{SCF} - \epsilon_{e_g}^{SCF} < \tfrac{1}{2}(J_{e_g} + J_{t_{2g}})$$

and since the e_g orbital in the new scheme is predominantly a metal $3d$ orbital, J_{e_g} and $J_{t_{2g}}$ are large.

The case of some rare-earth fluorides and oxides is included by *Jørgensen* in the third revolution in ligand field theory (*36, 38*) and there is little doubt that the ionization energies of the half-filled $4f$ orbitals in, for example, Gd(III) are larger than the ionization energies of the doubly occupied fluorine or oxygen $2p_\sigma$ orbitals. These compounds are, in fact, likely candidates for the paradoxical breakdown of Koopmans' theorem. Firstly, the interelectronic repulsion term in $4f$ orbitals is very high (approximately 24 eV for the $4f$ orbital in Gd(III) (*71*). Secondly the difference in the SCF eigenvalues of the mainly $4f$ bonding orbitals and the predominantly ligand σ-orbital is very small, as shown by the very weak nephelauxetic effect of the $4f$ orbitals in rare-earth complexes (*72*). It is true that the interelectronic repulsion parameter of the fluorine or oxygen σ-orbital is not small, but there is a third factor contributing to the paradoxical breakdown of Koopmans' theorem in these compounds. This is the Madelung stabilizing energy. If we assume again that the diagonal matrix element of the rare-earth metal $4f$ orbital is lower than that of the ligand $2p_\sigma$ orbital, the Madelung energy will strongly stabilize the slightly anti-bonding doubly occupied ligand orbital. The ionization energy of this orbital, that is, the value $\left(\dfrac{\delta \epsilon^{SCF}}{\delta n} \right)$ at the point $n = \tfrac{3}{2}$ is

$$I_{2p} - \sum_\nu n_\nu(\nu, 2p) - J_{2p,2p} + E_{\text{Mad}} .$$

In other words, to the Hartree-Fock potential one must add a Madelung term. Because of this extra term, the configuration $(\ldots (4f)^1 \, (2p)^2)$ is more stable than $(\ldots (4f)^2 \, (2p)^1)$ provided the following condition obtains:

$$\epsilon_{2p}^{\mathrm{SCF}} - \epsilon_{4f}^{\mathrm{SCF}} < \tfrac{1}{2}\{J_{4f,4f} + 2E_{\mathrm{Mad}} - J_{2p,2p}\} \qquad (22)$$

Jørgensen (*40–42*) has insisted that the main paradox derived from these results is that the system M(IV) lacking one $4f$ electron and the fluorine or oxygen systems lacking one $2p$ electron are almost degenerate in these compounds. In models of the type of Hückel's, this near equalization of the diagonal elements is a necessary condition for strong covalency, which seems to contradict the nephelauxetic evidence (*72*) that both the symmetry restricted and central-field covalencies are extremely weak. Strong covalency, however, needs not only approximately equal diagonal elements but also a large overlap, a condition that does exist between rare-earth metal $4f$ orbitals and ligand σ-orbitals. If, as we assumed, the $4f$ diagonal element is slightly lower than the ligand $2p_\sigma$ diagonal element, the bonding, predominantly $4f$, orbital remains half-filled, whereas the anti-bonding, predominantly $2p_\sigma$ orbitals, is doubly-occupied and the nephelauxetic ratio should remain close to unity.

It should finally be remarked that the multiplet splittings in the rare-earths are so large that molecular orbital concepts loose most of their meaning. Paradoxical violations of Koopmans' theorem are less likely to occur, and more difficulty to describe in such clear-cut terms, in compounds of the $4d$ and $5d$ transition elements and in the actinides. On one hand, the interelectronic repulsion terms of the $4d$, $5d$ and $5f$ orbitals are smaller than those of the $3d$ and $4f$ orbitals, and it is also generally true that bonding-antibonding separations tend to be larger. On the other hand, relativity effects become increasing important and it is the whole simple molecular orbital scheme that breaks down, rather than a single aspect such as Koopmans' theorem (*73*).

Acknowledgement: I am greatly in debt to Professor *C. K. Jørgensen* for the benefit of many helpful discussions during my stay at the University of Geneva.

References

1. *Strong, L. E.:* Foreword to Ida Freund's "The Study of Chemical Composition", Cambridge University Press, 1904; reprinted by Dover Publications, Inc., New York, 1968.
2. *Jørgensen, C. K.:* Topics Current Chem., *56*, 1 (1975).
3. *Dewar, M. J. S.:* "The Molecular Orbital Theory of Organic Chemistry", New York: McGraw-Hill Inc., 1969, p. 191.
4. *Jørgensen, C. K.:* Modern Aspects of Ligand Field Theory. Amsterdam: North-Holland Publishing Co., 1971, pp. 471–510.
5. *Linderberg, I., Öhrn, Y.:* Propagators in Quantum Chemistry. New York: Academic Press, 1973.
6. *Dewar, M. J. S.:* Chem. in Britain, *11*, 97 (1975).
7. *Pauling, L.:* In: Perpectives in Organic Chemistry, pp. 1–8 (A. Todd, Ed.), New York: Interscience, 1956. In this paper Pauling proffered the opinion that the resonance theory is an extension of the classical structural theory, rather than quantum-mechanical in character.
8. *Shull, H.:* J. Chem. Phys. *30*, 1405 (1959); Allen, T. L., Shull, H.: J. Chem. Phys., *35*, 1644 (1961).
9. *Levy, M., Stevens, W. J., Shull, H., Hagstrom, S.:* J. Chem. Phys., *52*, 5483 (1970).
10. *James, H. M., Coolidge, A. S.:* J. Chem. Phys., *1*, 825 (1933).
11. *Kekulé, A.:* Ann. Chem. Pharm., *106*, 129 (1858).
12. *Mulliken, R. S.:* Chem. Revs., *9*, 347 (1931); Rev. Mod. Phys., *4*, 1 (1932).
13. *Koopmans, T. A.:* Physica, *1*, 104 (1933).
14. *Jørgensen, C. K.:* ref. (*4*), pp. 98–104.
15. *Lorquet, J. C.:* Rev. Mod. Phys., *32*, 312 (1960).
16. This vivid expression was suggested to me by Professor Jørgensen.
17. *Bagus, P. S.:* Phys. Rev., *139*, A, 619 (1965).
18. *Löwdin, P. O.:* Adv. Chem. Phys., *2*, 207 (1959).
19. *Clementi, E.:* J. Chem. Phys., *38*, 2248 (1963), *39*, 175 (1963).
20. *Allen, L. C., Clementi, E., Gladney,:* Rev. Mod. Phys., *35*, 465 (1963).
21. *Siegbahn, K., Nordling, C., Johansson, G., Hedman, J., Hedén, P. F., Hamrin, K., Gelius, U., Bergmark, T., Werme, L. O., Manne, R., Baer, Y.:* ESCA Applied to Free Molecules. Amsterdam: North-Holland Publishing Co., 1969.
22. *Cade, P. E., Sales, K. D., Wahl, A. C.:* J. Chem. Phys., *44*, 1973 (1966).
23. *Connolly, J. W. D.:* Int. J. Quant. Chem., *6*, 201 (1972).
24. *Cederbaum, L. S., Hohlneicher, G., Niessen, W. von:* Chem. Phys. Letters, *18*, 503 (1973).
25. *Basch, H., Hollister, C., Moscowitz, J. W.:* Chem. Phys. Letters, *6*, 204 (1969).
26. *Demuynck, J., Veillard, A., Wahlgren, U.:* J. Am. Chem. Soc., *95*, 5563 (1973).
27. *Evans, S., Green, J. C., Green, M. H. L., Orchard, A. F., Turner, D. W.:* Disc. Faraday Soc., *47*, 112 (1969).
28. *Coutière, M. M., Demuynck, J., Veillard, A.:* Theoret. Chim. Acta, *27*, 281 (1972).
29. *Evans, S., Green, M. H. L., Jewitt, B., Orchard, A. F., Pygall, C. F.:* Trans. Faraday Soc. II, *68*, 1847 (1969); the 6.9 eV band was attributed to the e_{2g} electron and the band at 8.7 eV was assigned to the e_{1u} electron.
30. *Baerends, E. J., Ros, P.:* Chem. Phys. Letters, *23*, 391 (1973); SCF-X_α values: $I_{e_{2g}} = 6.7$ eV and $I_{e_{1u}} = 8.1$ eV.
31. *Hand, R. W., Hunt, W. J., Schaefer III, H. F.:* J. Am. Chem. Soc., *95*, 4517 (1973).
32. *Jørgensen, C. K.:* Oxidation Numbers and Oxidation States. Berlin–Heidelberg–New York.: Springer, 1969.
33. *Demuynck, J., Veillard, A.:* Theoret. Chim. Acta, *28*, 241 (1973).
34. *Evans, S., Hamnett, A., Orchard, A. F., Lloyd, D. R.:* Disc. Faraday Soc., *54*, 227 (1973).
35. *Wertheim, G. K., Rosencwaig, A., Cohen, R. L., Guggenheim:* Phys. Rev. Letters, *27*, 505 (1971).

36. *Jørgensen, C. K.:* Chimia, *25*, 213 (1971); ibid., *26*, 252 (1972).
37. *Fukui, K.:* Theory of Orientation and Stereoselection. Berlin–Heidelberg–New York: Springer, 1970.
38. *Jørgensen, C. K.:* Chimia, *27*, 203 (1973).
39. *Campagna, M., Bucher, E., Wertheim, G. K., Buchanan, D. N. E., Longinotti, L. C.:* Proc. 11th Rare-Earth Research Conf., Traverse City, Michigan, (1974), p. 810.
40. *Jørgensen, C. K.:* Structure and Bonding, Vol. *22,* p. 49. Berlin–Heidelberg–New York. Springer 1975.
41. *Jørgensen, C. K.:* Chimia, *28*, 6 (1974).
42. *Jørgensen, C. K.:* Adv. Quant. Chem., *8*, 137 (1974).
43. *Héden, P. O., Löfgren, H., Hagström, S. B. M.:* Phys. Rev. Letters, *26*, 432 (1971).
44. *Hagström, S. B. M., Brodén, G., Héden, P. O., Löfgren, H.:* J. Physique (Colloque CNRS no. 196) C 4–269 (1971).
45. *Cox, P. A., Baer, Y., Jørgensen, C. K.:* Chem. Phys. Letters, *22*, 433 (1973).
46. *Baer, Y., Busch, G.:* J. Electron Spectroscopy, *5*, 611 (1974).
47. *Dewar, M. J. S., Hashmall, J. A., Venier, C. G.:* J. Am. Chem. Soc., *90*, 1953 (1968).
48. *Kollmar, H.:* Chem. Phys. Letters, *8*, 533 (1971).
49. *Dewar, M. J. S., Kollmar, H., Suck, A.:* Theoret. Chim. Acta, *36*, 125 (1975).
50. *Roothaan, C. C. J.:* Rev. Mod. Phys., *32*, 179 (1960).
51. *Slater, J. C.:* Int. J. Quant. Chem., *S3*, 727 (1970); Adv. Quant. Chem., *6*, 1 (1972).
52. *Slater, J. C., Wood, J. H.:* Int. J. Quant. Chem., *94*, 3 (1971).
53. *Johnson, K. H., Smith Jr., F. C.:* Phys. Rev., *B5*, 831 (1972).
54. *Slater, J. C., Johnson, J. H.:* Phys. Rev., *B5*, 844 (1972).
55. *Jørgensen, C. K.:* Orbitals in Atoms and Molecules. London: Academic Press 1962.
56. *Klopman, G.:* J. Am. Chem. Soc., *86*, 1463, 4550 (1964); J. Chem. Phys., *43*, S 124 (1965).
57. *Baird, N. C., Whitehead, M. A., Sichel, J. M.:* Theoret. Chim. Acta, *11*, 38 (1968).
58. *Ferreira, R., Bates, J. K.:* Theoret. Chim. Acta, *16*, 111 (1970).
59. The method is implicit in a paper by *Longuet-Higgins, H. C., Pople, J. A.:* Proc. Phys. Soc., *68A*, 591 (1955).
60. *Ferreira, R.:* J. Chem. Phys., *49*, 2456 (1968).
61. By "correct" we mean equal to that given by *Roothaan*'s method (Ref. *50*).
62. *Mulliken, R. S.:* J. Chem. Phys., *2*, 782 (1934); *3*, 586 (1935).
63. *Iczkowski, R. P., Margrave, J. L.:* J. Am. Chem. Soc., *83*, 3547 (1961).
64. *Hinze, J., Whitehead, M. A., Jaffe, H. H.:* J. Am. Chem. Soc., *85*, 148 (1963).
65. *Ferreira, R.:* Adv. Chem. Phys., *13*, 55 (1967).
66. *Brundle, C. R., Robin, M. B., Kuebler, N. A., Basch, H.:* J. Am. Chem. Soc., *94*, 1451 (1972).
67. *Jørgensen, C. K.:* Chimica Teorica, VIII Corso Estivo di Chimica, Milano, 1963, p. 63. Rome: Academia dei Lincei, 1965.
68. See, for example, the excellent book by *Schläfer, H. L., Gliemann, G.:* Basic Principles of Ligand Field Theory (translated by D. F. Ilten). London: Wiley-Interscience 1969, p. 105.
69. For example, the J_{μ_μ} integral for the highest occupied σ-orbital of the hexafluoroacetylacetonate ligand is small rather than large.
70. *Hillier, J. H., Saunders, V. R.:* Proc. Roy. Soc., *A230*, 161 (1970).
71. *Jørgensen, C. K.:* Structure and Bonding, Vol. *13*, p. 199, Berlin–Heidelberg–New York 1973.
72. *Jørgensen, C. K.;* Prog. Inorg. Chem., *4*, 73 (1962).
73. Quite recently it was shown (Giambiagi, M., Giambiagi, M. S., Ferreira, R., Blanck, S.: Chem. Phys. Letters, *38*, 65 (1976)) that in *trans*-butadiene the derivatives of a given eigenvalue with respect to the occupation numbers of the other eigenfunctions are second-order terms when compared with the derivative of the eigen-value with respect to its own occupation number.

Band and Localized States in Metallic Thorium, Uranium and Plutonium, and in Some Compounds, Studied by X-Ray Spectroscopy

C. Bonnelle

Laboratoire de Chimie Physique, 11, rue Pierre et Marie Curie, 75005 Paris, France.

Table of Contents

1. Introduction

Actinides have special properties that are due to the presence of the partially filled $5f$ shell. Their $6d$ and $7s$ electrons form the conduction band in the metal whose energy is close to the $5f$ binding energy. Differing behaviours have been found between these elements and the rare earths which are characterized by the filling of the $4f$ shell. The rare earth $4f$ electrons are highly localized: they determine the magnetic properties in the metal and have little effect on chemical behaviour and on the physical properties that arise from conduction or valence distributions. Except for cerium, which is tetravalent in its stable compounds and has particular properties as a function of temperature and pressure (1), and for europium and ytterbium, which are divalent in the metal and certain compounds, the rare earths have three valence electrons throughout the series and have an integral number of $4f$ electrons (2).

On the contrary, the valence of the actinides varies and there is some uncertainty as to the configuration in the metals. In particular, in the first half of the series, the number of $5f$ electrons can vary with temperature and pressure and is not an integer. Moreover, the magnetism is present in the metals only from the middle of the series (3). On the other hand, in the case of ionic and covalent compounds, a strong intra-atomic correlation between electrons from the $5f$ states has been found and the experimental data clearly show that the $5f$ states have an atomic-like character (4); indeed, the $5f$ electrons in the actinide compounds present a magnetic behaviour which is similar to that of the $4f$ electrons in the rare earth compounds.

Then, the $5f$ electrons in actinides appear less localized than the $4f$ electrons in rare earths. Moreover, their behaviour is different from that of d electrons in the transition metals. The $5f$ electrons therefore constitute an intermediate group of electrons; in order to understand their behaviour, it is thus necessary to take into account properties involving localized or delocalized electrons.

In the past few years, theoretical studies of band structures have been developed with the aim of explaining the particular electronic properties of actinide metals (5). Various models have been proposed to describe these metals. They emphasize the importance of correlations on the determination of the localized or non-localized character of the $5f$ electrons. These correlations depend on the overlap between the $5f$ orbitals belonging to neighbouring atoms and, therefore, on the separation of the actinide ions in the solid.

It is only more recently that studies by X-ray spectroscopy have been undertaken for the actinide metals and compounds. Simultaneously, the X-ray spectroscopy of the rare earths has been widely developed and a comparison is possible to point out the differences and similarities between the actinides and the lanthanides.

We present here the data on the electronic distributions in Th, U, and Pu metal and oxide that can be obtained from X-ray spectroscopy. Firstly, we give a brief summary of the mechanisms which occur in X-ray emission and absorption spectra for solids having a localized or unlocalized incomplete subshell. Then we discuss typical information obtained by this method for the rare earths, particularly for their $4f$ distribution. Lastly, we present our experimental results for the actinides with a discussion on the character of $5f$ filled and empty states in the metals and stable oxides.

2. Principles of X-UV Spectroscopy

For many years, X-UV emission and absorption spectroscopy has been known as an appropriate method for the determination of electronic distributions in solids. An analysis of X-ray emission spectra that are due to transitions between an inner nl atomic level of known characteristics and the various levels either in the energy bands or in the molecular orbitals provides a direct method of determining the distribution of occupied conduction or valence states in the material to be studied. Absorption spectra on the other hand describe the photoabsorption process and provide a description of unoccupied states over a wide energy range. These two types of analysis are complementary.

The intensity is governed by the dipolar transition probabilities and can be calculated by use of the golden rule (6). It is possible to obtain a simple description with the help of the optical selection rules; according to these, the transition probabilities towards the $(l + 1)$ states are strong. The analysis of transitions involving inner shells of various l numbers allows one to determine the outer occupied or unoccupied states of various symmetries separately, even though their energies overlap. The method has been used with success to determine the density of s or p states of light elements, in metals or insulators, and the d distribution of transition metals and their compounds, ... In the case of the rare earths and actinides, the f electron distributions can be deduced from $n\,d_{3/2}$ and $n\,d_{5/2}$ spectra and the $d-s$ electron distributions from the $n\,p_{1/2}$ and $n\,p_{3/2}$ spectra. The number n can be chosen so that the spin orbit coupling is large enough to obtain a clear energy separation of the transitions corresponding to the two j values.

In fact, it is necessary to take into account the finite life time of the inner hole and the spectrum is the convolution product of the outer electronic distribution and the Lorentz distribution of the inner atomic level (7). As a consequence, the resolution depends on two factors: one is the instrumental broadening; the other is intrinsic and is due to the width of the inner level.

The role of many body effects must also be considered. Many studies have been made and it has been concluded that the perturbation which follows the sudden creation of a core hole is generally weak on the outer electronic distribution provided that the solid in its fundamental state has no component with an outer incomplete localized shell. In metals, as a consequence of the presence of the hole, low-energy-conduction electron-hole pairs are created (8), leading to an anomaly near the Fermi level and to a small asymmetric broadening of the emission band (9). For semiconductors or insulators, excitonic transitions have been predicted but have not been clearly observed except perhaps in alkali halides (10).

Lastly, the X-UV emission and absorption spectra depend on the chemical bonding and information on the oxidation state, the hybridization, the fractional charges, ... can be obtained from these.

If the elements present in the solid have an incomplete strongly localized shell, the spectra show special features. This is the case, for example, for the rare earths, and we shall discuss this point successively for the emission and for the absorption processes.

Emission. Generally, the emission lines are emitted after an *ionization* in an inner atomic shell; an electronic reorganisation follows immediately after the creation of the inner hole and radiative transitions are observed at lower energies than the absorption discontinuity.

Recently, lines of a new type have been observed in the rare earth X-ray spectra, after an *excitation*: the initial state of this transition is a highly excited neutral atom where a $3d$ electron is promoted to one of the empty f states depending on the dipolar transition probabilities. This electron can come back to its initial $3d$ level by a monoelectronic radiative transition. These emissions are in coincidence with a strong absorption peak and are interpreted as a resonance radiation (11). We have labeled these resonance lines (or R lines). They correspond to the $3d^9 4f^{n+1} \rightarrow 3d^{10} 4f^n$ radiative transition which is the reverse of absorption.

The excited initial state can be produced either directly or by means of radiative cascades and non-radiative transfers. The first is the most probable process, i.e. a filling up by direct inner shell excitation. As a consequence, the R lines are observed only when the hole is created by electron bombardment and not by bremsstrahlung irradiation (12).

For resonance lines to be observed, the excitation process must be large enough with respect to the ionization process; it must correspond to a dipolar transition without change of spin in order that the radiative deexcitation may have a large probability. Moreover, it is necessary that the excited electron does not diffuse in the solid during the life of the inner hole (13). So that diffusion does not have time to take place, the interactions with the surrounding atoms must remain weak; therefore, the excited state must be localized with regard to the time scale of the experiment which depends on the hole. The latter condition is very restrictive. It is not satisfied if the excited state belongs to a continuum in the solid or is a discrete state susceptible to have a rapid interaction by an autoionization process with a continuum state of the same energy, same J, and same parity. In fact, in these cases, the excited state normally goes to $3d^9 4f^n$ before the radiative resonance transition can be emitted. Then, for a R line to be emitted, the life time of the excited state must depend essentially on the life time of the inner nl hole and the deexcitation must almost exclusively take place from transitions to this hole.

Indeed, competing non-radiative processes are present in the decay of the excited state. They correspond to Auger transitions in the excited atom and high energy Auger lines appear in the corresponding *nl* spectrum (*14*).

Absorption. During the X-UV photoabsorption process, an electron of a more or less deep atomic level is excited to unoccupied states of suitable symmetry.

When the unoccupied states are itinerant, a marked increase in the variation of photoabsorption coefficient forming a discontinuity appears at an energy just sufficient for the transfer of an electron to the first empty levels. The inflexion point of the discontinuity corresponds to the position of the Fermi level in a metal or the bottom of the conduction band in a semi-conductor or an insulator. The ratio between the photoabsorption coefficient on either side of the discontinuity is called the absorption jump. If the density of states is uniform, the shape of the discontinuity is that of the arctangent curve. When a high density of unoccupied states of the appropriate symmetry is situated near the Fermi level, an absorption maximum can be expected.

In the case of localized empty states one or more absorption lines are observed in the variation of the photoabsorption coefficient. Beyond this an absorption jump is generally observed; it corresponds to the transitions toward hybridized continuum states of positive energy and its inflexion point gives the ionisation energy.

For the solids without outer incomplete localized shells, the many-body effects are generally treated by considering the life time of a core hole as very long with respect to the electron-electron processes. This condition is not verified when an incomplete localized shell is present, i.e. resonance lines can be observed and the problem must be considered differently. If the overlap between the wave-functions of photo-hole and holes in the incomplete localized shell is large enough, the exchange interaction is noticeable. Then, the absorption spectrum has a multiplet structure but only the transitions satisfying the selection rule $\Delta J = 0, \pm 1$, from the fundamental state are observed (*15*). These various structures are found also more or less strongly in the resonant part of the emission spectrum.

More generally, the formation of an inner hole is accompanied by a polarization effect which produces an increase of the binding energy of about the same order of magnitude for all atomic levels. The polarization takes place rapidly and a modification in the energy of transition between two localized levels could be observed but it should remain weak.

In an excitation process, the electron and the hole can remain bound, producing an exciton state just below the conduction band. Indeed, the mass of an inner hole is considered as infinite and the exciton binding energy is thus almost zero with reference to the absorption threshold energy. If the resonance lines were excitonic type transitions, the emission spectrum should be exactly the reverse of absorption. We would see that this is not the case: although a localized excited M_{IV} state has a large probability of existing, sometimes the resonance M_{IV} lines are absent, whereas the resonance M_V lines are the most intense of the spectrum (*11*).

3. Electronic Distributions of Rare Earths

In this section, the results which can be obtained from $3d$ and $3p$ rare earth spectra are discussed. The spin-orbit splitting is large enough so that the $3d_{3/2}$ and $3d_{5/2}$ spectra are clearly separated; this is also true for the $3p_{1/2}$ and $3p_{3/2}$ spectra. On the other hand, this is not the case for the $4d$ and $4p$ spectra.

$4f$ *distribution of metals.* In principle, the observed emission intensity, I, which corresponds to the monoelectronic transitions from $4f$ states toward the initially ionized $3d$ shell, ($M\alpha$ and $M\beta$ lines) gives the $4f^n$ distributions. A process of the Auger type can take place at the end of the emission, one conduction electron going to fill the $4f$ hole, and another conduction electron being excited in order that the energy conservation be satisfied (*11*). This transition is shown schematically in Fig. 1. Its importance depends on the respective values of the $4f$ hole lifetime and of the Auger recombination time. The further below the Fermi level is the $4f^n$ level the larger is the probability of the process.

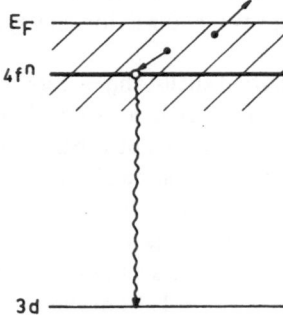

Fig. 1. Principle of the $4f^n$ transitions.

The presence of such a process can explain the characteristics of $M\alpha$ and $M\beta$ lines: these emissions behave like emission bands; they are asymmetrical in the metal, clearly broader than the absorption lines, and their position and shape vary with binding. Their broadening is due to fast collisions of the conduction or valence electrons with the $4f$ hole and depends on these electron distributions, explaining that the $M\alpha$ and $M\beta$ shape is different in the metal and the oxide. Lastly, the energies of $M\alpha$ and $M\beta$ are not characteristic of the atomic transition $3d^9\,4f^n \rightarrow 3d^{10}\,4f^{n-1}$. The lifetime of the $4f$ hole is very short in the metal and the final state of the emission has a strong probability of corresponding to the $4f^n$ configuration.

The excited $4f^{n+1}$ state distributions can be deduced from the variation of the linear photoabsorption coefficient μ accompanying the ejection of a $3d_{3/2}$ or $3d_{5/2}$ electron. In fact, because of the existence of the incomplete $4f$ shell, a strong absorption to the $4f$ empty states is expected in the $3d$ spectra. A process of the Auger type analogous to that described for the emission should be possible, but its probability is very weak; during this process, the excited electron could fall to an empty state of the conduction band, a metal conduction electron being excited for energy conservation.

Indeed, it has been shown that an electron ejected into a $4f^{n+1}$ orbital has a high degree of localization which is due to the penetration of the centrifugal potential barrier by the $4f$ orbitals (16). Thus, within the independent particle model, the most probable way of ionizing a nd electron should be to excite it to a $4f$-orbit from which it could escape to continuum f-orbits and, to a lesser extent, continuum p-orbits. As a consequence, absorption lines corresponding to large photoabsorption cross-sections are present in the $3d$ spectra. Moreover, it is remarkable that no jump is observable in the $3d_{5/2}$ spectra of the rare earths, while in the $3d_{3/2}$ spectra, a very small jump appears only for some heavy rare earths (17).

The first condition for the observation of a resonance line has been satisfied. Intense resonance lines have indeed been observed in the M_{IV-V} emission spectra. Their size shows that an appreciable amount of overlap exists between the $4f^{n+1}$ orbital and the $3d$ shell. Moreover, we have seen that these lines are observable only if the radiative rate is sufficient compared with other processes of de-excitation to ensure that the excited state has a noticeable probability of depopulation by direct radiative decay. In particular, the diffusion of the $4f$ supplementary electron to continuum states must remain weak; this condition is entirely compatible with the relatively large localization of the $4f$ rare earths states mentioned above.

Finally, the X-ray emission spectra reveal that the $3d^9\ 4f^{n+1}$ state is strongly resonant. Its population is governed by two opposite effects: a very short lifetime and a high electron impact excitation rate. Because of the strong reabsorption of resonance radiation, the "effective" lifetime may be almost comparable to the lifetime of nonresonant $3d^9\ 4f^n$ states.

The size of the resonance lines is different from that of the absorption lines: for example, the M_{IV} absorption of lanthanum is larger than the M_V absorption, whereas the M_{IV} resonance line is less intense than the M_V line (18). For heavy rare earths, only M_V resonance lines have been observed up to now; however, the total intensities of the M_{IV} and M_V emission spectra remain in the ratio of the statistical weights because $M\beta$ is much more intense than $M\alpha$ (19). Spin relaxation effects different for $3d_{3/2}$ or $3d_{5/2}$ excited electrons could be invoked to explain these measurements.

We have seen that the inner hole can produce a perturbation of the electronic distribution relatively to that of the unperturbed solid. Various analyses have shown the importance of the final-state configuration on the spectra and the electron-hole interactions which can, in some cases, alter their shape. Thus, the rare earth $3d$ photoabsorption spectra present a number of structures which spread over several eV. They can be interpreted as the components of multiplets because of the exchange coupling between the $3d$ and $4f$ shells. The perturbation weakens as the hole lies in a deeper inner shell. In fact, the exchange interaction strength depends on the overlap between the wavefunctions of the inner hole and the localized $4f$ holes; the weaker

29

the overlap, the smaller the exchange interaction. It is clearly smaller than the spin-orbit interaction for the $3d$ spectra. The M_{IV-V} absorption spectra of various metals, lanthanum, cerium (20), and more recently europium and gadolinium (21), have been well interpreted in terms of $3d^{10} 4f^n \rightarrow 3d^9 4f^{n+1}$ transitions in the ions of $4f^0, 4f^1$ and $4f^7$ initial configurations, respectively.

4f-Fermi level distances. The conduction or valence $5d\,6s$ distributions can be deduced from the M_{III} spectra which involve the internal atomic p levels. For a metal, the inflexion point of the M_{III} absorption discontinuity corresponds to the position of the Fermi level and the emission and absorption spectra overlap at the Fermi level. For an insulator, a forbidden band appears between the emission and absorption spectra; this is the case, for example, of Eu_2O_3 which has a gap of about 3 eV (22). The inflexion point of the M_{III} discontinuity which corresponds to the transitions to the bottom of the conduction band is generally situated several eV toward the higher energies with respect to the metal.

The position of the occupied $4f$ states with respect to the Fermi level can be determined by taking into account the M_{III} and M_{IV-V} spectra and also various X-ray atomic emission lines which can give the energy differences $(E_{M_{III}} - E_{M_{IV}})$ or $(E_{M_{III}} - E_{M_V})$. In fact, the energy difference D between the $4f^n$ state and the Fermi level is

$$D = E_F - [(E_{M_{III}} - E_{M_{IV}}) + E_{M\beta}]$$
$$= E_F - [(E_{M_{III}} - E_{M_V}) + E_{M\alpha}]$$

where E_F is the energy of the inflexion point of the M_{III} absorption curve (23). This method supposes that the perturbations due to $3p$ or $3d$ holes are comparable even for different outer distributions. It may be argued that the $M\alpha$ or $M\beta$ emissions give the $4f^n$ configuration only, because Auger type recombinations are present at the emission final state. If this process does not exist, the configuration should be $4f^{n-1}$ as in photoemission (24).

4f distribution of compounds. A displacement of the M_{IV} and M_V absorption spectra is observed when the electronic configuration of the rare earth varies. For Europium (22), for example, the absorption spectra of Eu in Eu_2O_3 are displaced by 3.0 eV toward the higher energies with respect to EuO. This displacement corresponds to the energy difference between the $3d^9 4f^7$ and $3d^9 4f^8$ configurations which are associated with the absorption transition final states of each compound, respectively. The displacement is accompanied by a modification in the absorption structures following the change in the electronic configuration. This is due to different demultiplications of the energy levels involved in the absorption transitions. The resonance lines shift in the same way as the corresponding absorption lines. The shift of $M\alpha$ and $M\beta$ is 0.3 eV toward the higher energies. Analogous modifications are observed for the other divalent rare earth, Yb, between the metal and the oxide, Yb_2O_3 (17).

For the trivalent rare earths, the number of $4f$ electrons remains the same in the metal and in the oxide. Therefore, one should not observe a modification in the d absorption spectra. However, a modification of the shape and width of the $M\alpha$ and $M\beta$ emissions appears with the chemical binding. For instance, for gadolinium, the emission shift is about 1.0 eV toward the lower energy in the oxide and the $M\alpha$ and $M\beta$ shape is modified; the emissions are symmetrical in the oxide and not in the metal (11). To explain these results, we have seen that it is necessary to take into account the secondary effects of the Auger type arising from the transition of conduction or valence electrons toward the $4f$ hole present in the final state of emission. It is also possible to consider, in the metal, the exchange interaction between the $4f$ electrons and the conduction electrons. The exchange energy is 0.6 eV for gadolinium, thus of the same magnitude as the observed emission shifts (25).

The oxide CeO_2 constitutes a particular case: in fact, the interpretation of M_{IV-V} absorptions in terms of the $3d^{10} 4f^n \rightarrow 3d^9 4f^{n+1}$ atomic transition with $n = 0$ or 1 is not possible. However, in Ce metal, we have found a good agreement with this transition for $n = 1$. The CeO_2 M_{IV-V} absorptions are displaced by about 1.5 eV toward the higher energies with respect to metal; this displacement suggests an increase in the oxidation number. On the other hand, the M_{IV-V} emission spectra reveal the presence of $4f$-type occupied states in CeO_2. These states must be strongly mixed with $2p$ oxygen orbitals and these results can be explained by invoking the partially covalent character of the bonding.

A modification of the electronic structure with decreasing grain size has been observed for small ytterbium particles of about 30 Å diameter; the promotion of one $4f$ electron into the conduction band is shown from the M_{IV-V} emission spectra. It is suggested that the same modification must be expected for all the elements having a strongly localized shell situated a little below the Fermi level (26).

4. Electronic Distributions of Th, U, and Pu

We report here the principal results obtained from the analysis of M_{IV} and M_V emission and absorption spectra and also of the M_{III} absorption, for Th, U, and Pu. Many difficulties have had to be surmounted to obtain clean metal and oxide samples. The metal samples were obtained by successive evaporations of Al (or SiO), the actinide, and again Al (SiO), forming protected "sandwich" layers of suitable thickness. Their purity is controlled by electron or X-ray diffraction (27).

Fig. 2 shows a diagram summarizing the various transitions which can be observed in the M_{III} and M_V spectra of a metal as well as in the $3d$ Auger spectra. The M_{III} and M_V absorption transitions are shown in Fig. 2a and b; the energy of the M_{III} discontinuity corresponds to the transfer of an inner $3p_{3/2}$ electron to the Fermi level and its shape involves the $6d$ unoccupied distribution; the energy of the M_V absorption line is exactly that of the $3d^6_{5/2} 5f^n \rightarrow 3d^5_{5/2} 5f^{n+1}$ excitation transition. The M_V emission is shown in Fig. 2e: an inner $3d_{5/2}$ hole is created and a $5f$ electron transits to this hole with the emission of a photon. In the corresponding non-radiative transition, there is simultaneously the $5f$ electron transition, and the excitation or ionization of a $5f$ electron (or $6p$ or $6s$) (Fig. 2f). The M_V resonance line is represented in 2c: the excited $5f$ electron drops back to the inner hole; the corresponding emission line then coincides with an absorption line. The competing non-radiative transition is shown in 2d: this is an Auger transition in the excited atom; the final state has only one hole in an outer shell and the configuration is the same as in a photoemission process.

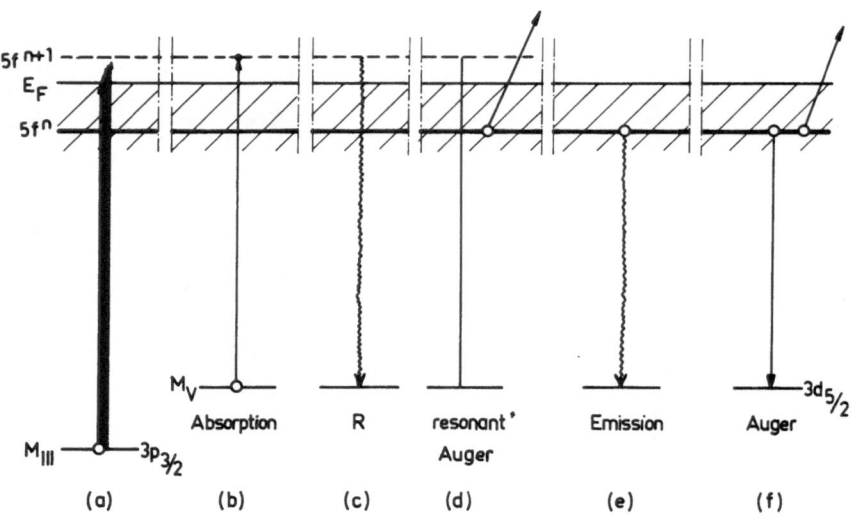

Fig. 2. Diagram of transitions from or to the M_{III} and M_V inner levels in the actinides.

4.1. Metals

In this paragraph we present the results obtained from the M_{IV-V} spectra for the $5f$ distributions of three metals in their stable phase at room temperature. Then we report the modifications observed for the $5f$ states of Pu in its δ phase. The results deduced from the M_{III} spectra are mentioned briefly and an energy diagram is proposed for each metal.

5f distributions. The M_{IV-V} emission and absorption spectra of uranium metal are presented in Fig. 3 (*28*). The emissions are excited by electron bombardment under voltage differences of between 1.2 to 2.5 times the ionization threshold. A resonance line is observed in the M_{IV} emission spectrum and also in the M_V spectrum, but only at a low excitation voltage. The M_V absorption line is about two times more

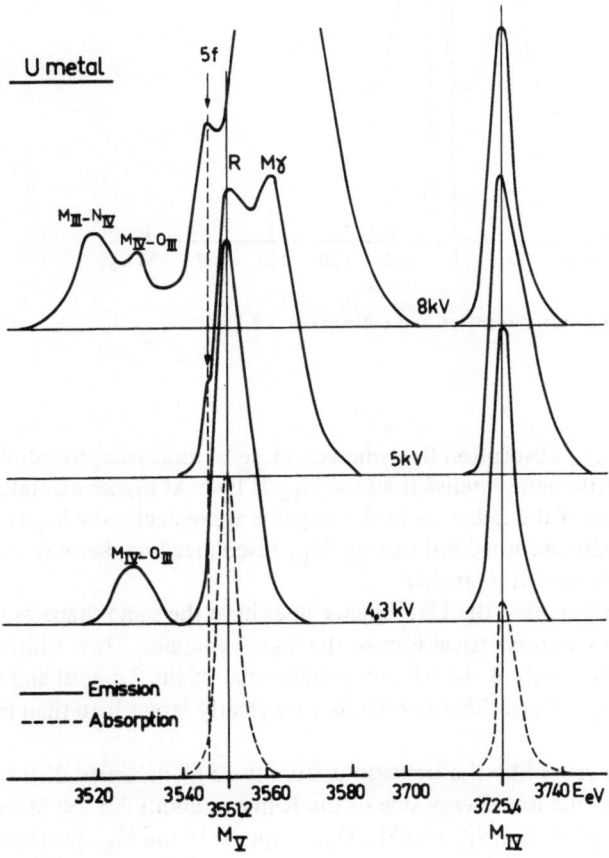

Fig. 3. M_{IV-V} α-U spectra: absorption and emission at various excitation voltages.

33

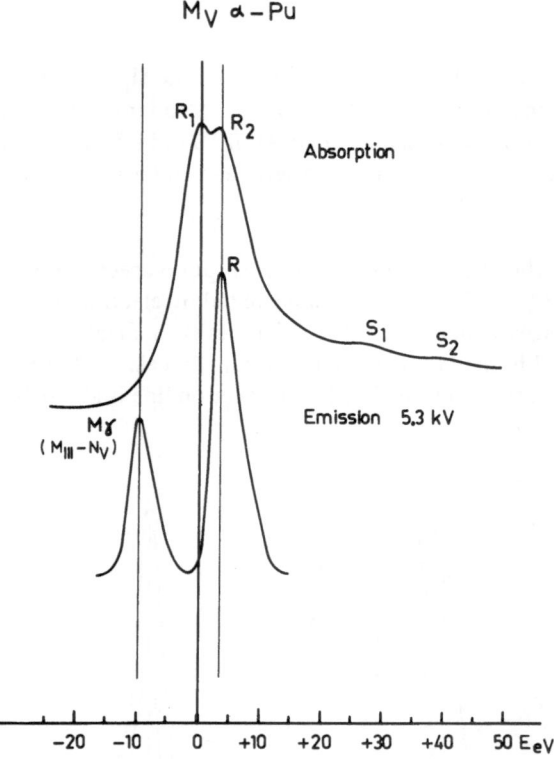

Fig. 4. M_V α–Pu spectra: absorption and emission at 5.3 kV.

intense than the M_{IV} absorption line whereas, close to ionization threshold, the M_V R line is only a little more intense than the M_{IV} R line. At higher excitation voltages, the self-absorption of the radiation in the target is appreciable; the M_V resonance line is almost totally absorbed but not the M_{IV} resonance line, because in this case the absorption coefficient is smaller.

As for the lanthanides, the U-resonance lines have the same shape as the absorption lines. They are asymmetrical toward the higher energies. Their widths involve simultaneously the width of the $5f$ empty state, that of the $3d$ level and the instrumental broadening effects; these two factors are clearly larger here than in the $3d$ rare earth spectra.

The emission caused by the transitions from the $5f^n$ occupied distribution in the ionized atom is on the low energy side of the R line at about 3.5 eV. Moreover, atomic lines – $M_{III}N_V(M\gamma)$, $M_{III}N_{IV}$ and $M_{IV}O_{III}$ – appear in the M_V spectrum; their intensity increases strongly with increased voltage and their presence perturbs the observations.

34

Resonance lines are also observed in the M_{IV-V} emission spectra of thorium but no line appears toward the lower energies, showing that no $5f$ electron is present in the unperturbed metal (29). The M_{IV} and M_V absorption lines are of comparable intensity; this is also true of the M_{IV} and M_V R lines, and consequently the M_V R line is always observed whatever is the acceleration voltage.

The variation in the emitted intensity as a function of the photon energy near the Pu M_V ionization limit is plotted in Figs. 4 and 5 compared to the variation of the photoabsorption coefficient (30). At 5.3 eV, the self-absorption effect remains relatively weak. The atomic line $- M_{III} N_V (M\gamma) -$ is situated toward the lower energies of the spectrum and perturbes this at high voltages. A resonance line is observed in coincidence with the one of absorption peaks, R_2. Its intensity decreases with increasing voltage because of the self-absorption effect. No resonance line appears at the energy of the R_1 absorption peak. The latter is at 2.5 eV of the R_2 peak.

The emission caused by the transitions from normally occupied $5f$ states in the non-excited metal must be situated on the low energy side of the absorption peaks. Its intensity must follow that of the $M\gamma$ atomic line and its observation is made difficult because the presence of this strong line.

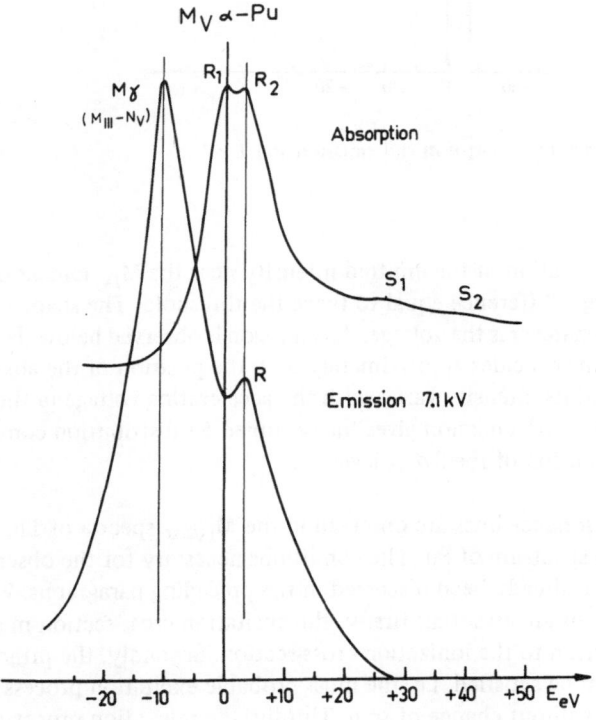

Fig. 5. M_V α–Pu spectra: absorption and emission at 7.1 kV.

35

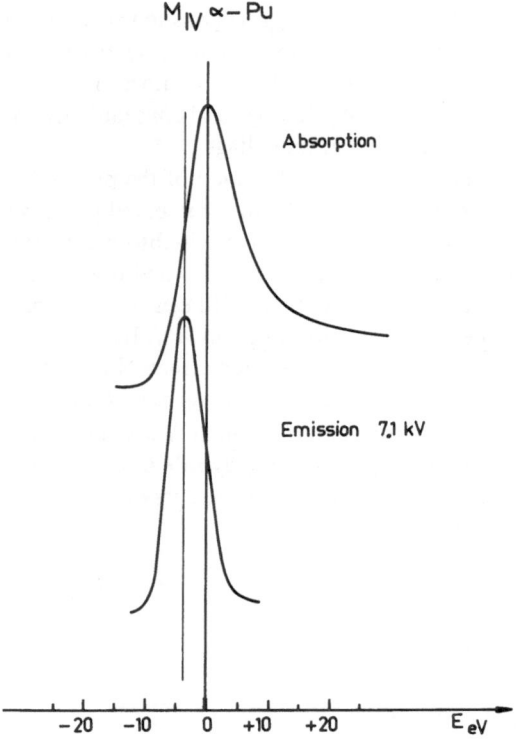

M_{IV} α – Pu

Absorption

Emission   7.1 kV

-20 -10 0 $+10$ $+20$ E_{eV}

Fig. 6. M_{IV} α–Pu spectra: absorption and emission at 7.1 kV.

In Fig. 6 the variation of the emitted intensity near the M_{IV} ionization limit is plotted for a voltage difference equal to twice the threshold. The shape of the spectrum is the same whatever is the voltage. An emission is observed below the absorption peak; its maximum coincides approximately with the position of the absorption curve inflexion point and its intensity varies with the accelerating voltage in the same way as does atomic line. This emission gives the occupied $5f$ distribution convoluted with the Lorentz distribution of the $3d_{3/2}$ level.

Intense $3d$ resonance lines are observed in the M_{IV-V} spectra of Th, U, and Pu, except in the M_{IV} spectrum of Pu. The conditions necessary for the observation of a resonance line have already been discussed in the preceding paragraphs. We have seen that three conditions are essential: firstly, the excitation cross-section must be large enough in comparison to the ionization cross-section. Secondly, the principle of the reverse return must be satisfied, i.e. the most probable excitation process must be a dipolar transition without change of spin. Thirdly, the relaxation processes not involving the inner hole decay must take place over a longer period than the lifetime of this hole. The latter constitutes a very restrictive condition in solids. This explains why, so far, the resonance lines have been observed only in a small number of cases. We have seen them in the $3d$ spectra of lanthanide metals and their compounds and also, with a very weak, intensity, in the $2p_{3/2}$ spectrum of Ni in NiF_2 but not in the $2p$ spectra of transition metals nor of Ni in NiO (31).

On the other hand, resonance lines have been observed in the K and $L_{II,III}$ spectra of molecular gases (32, 33). In these cases, the first two conditions do not seem to be fulfilled more satisfactorily than in the $2p$ spectra of transition elements, but the localization of the excited states must be greater.

It is thus possible to emphasize the following points: When the excited electron jumps in a wide conduction band, i.e. in itinerant states, the electronic speed in the band is sufficiently large for the electron to have a very small probability of returning to the inner hole. No resonance line is observed; this is the general case in the solid. On the other hand, if the excited state is strongly localized, i.e. if the electron stays during the emission process on the same atom with the same spin, a resonance line can be emitted. The existence of an exciton-type state could be taken into consideration to explain the presence of the resonance lines; but, except for thorium, the ratio of M_{IV} and M_V R line intensities is different from the ratio of the corresponding absorption line heights. In particular, the M_{IV} R line is absent in the spectrum of plutonium. As a consequence, the emission spectra cannot be interpreted suitably by excitonic transitions.

Then, from soft X-ray spectroscopy, it results that the excited $2p^5 3d^{n+1}$ states in the transition metals and in nickel oxide have a very short relaxation time to the ionized $2p^5 3d^n$ states, whereas the excited $3d^9 4f^{n+1}$ states in the lanthanides and $3d^9 5f^{n+1}$ in Th, U, Pu . . . are strongly resonant. As a consequence, the $4f$ and $5f$ empty states in the lanthanides and also in U, Pu, . . . cannot be described by a simple band model but must be considered as localized relative to the time scale of our experiments, i.e. 10^{-16} to 10^{-15} seconds. By analogy with the rare earths, it is possible to suggest the presence of a potential barrier hindering the excitation of a $3d$ electron towards higher energy orbits than the $5f$-orbit and thus hindering direct ionization. However, for the actinides in their phase stable at room temperature, both a jump and also structures are observed in the M_{IV-V} absorption spectra (see Fig. 4 and 7, structures S_1 and S_2). This point will be discussed in the next section.

The non-radiative decay process corresponding to the resonance radiation must contribute to the Auger spectra, This is a simple two-step process, namely excitation of an nd electron into the $5f$ shell, followed by an autoionizing transition i.e. $nd^9 5f^{n+1} \rightarrow nd^{10} 5f^{n-1} + e^-$. We suggest that the high energy features observed in the Auger $5d$ spectra of Pu (34), at 103 and 110 eV, be interpreted as caused by an Auger process from the excited $5d_{5/2}^{-1} 5f^{n+1}$ and $5d_{3/2}^{-1} 5f^{n+1}$ states, respectively.

The occuring of the resonance line in the $3d$ emission spectra suggests an appreciable amount of overlap between the empty $5f$ and $3d$ wave functions. As a consequence, the exchange interaction between the photohole and the $5f$ holes can be noticeable. Then, the final-state interactions should produce a demultiplication of photoabsorption spectra, analogous to that observed in the rare earth $3d$ spectra. Indeed, the M_V spectrum of Pu shows two peaks, R_1 and R_2, but only one of them has a corresponding resonance line. Thus, we suggest that these two peaks involve the presence of empty $5f$ states with different characteristics: hybridization, relaxation time, or other parameters (see also next section).

$5f$ *distribution of δ-Pu.* The plutonium $3d$ photoabsorption has been analyzed for modifications in the $5f$ excited states between the α and δ phases (*27, 35*). The same sample is maintained at 400 °C during the analysis to give the δ phase spectrum and at room temperature for the α phase. The variation of the photoabsorption coefficient with photon energy near the M_V ionization limit is plotted in Fig. 7. A very marked modification in the M_V photoabsorption according to the temperature is observed.

One line appears in the δ-Pu spectrum; it is asymmetrical toward higher energies: it is displaced by about -0.8 eV with respect to R_1 for α-Pu. Its half-width at half maximum, on the lower energy side, is about 2.3 eV. The shape of the $\delta-$Pu M_V ab-

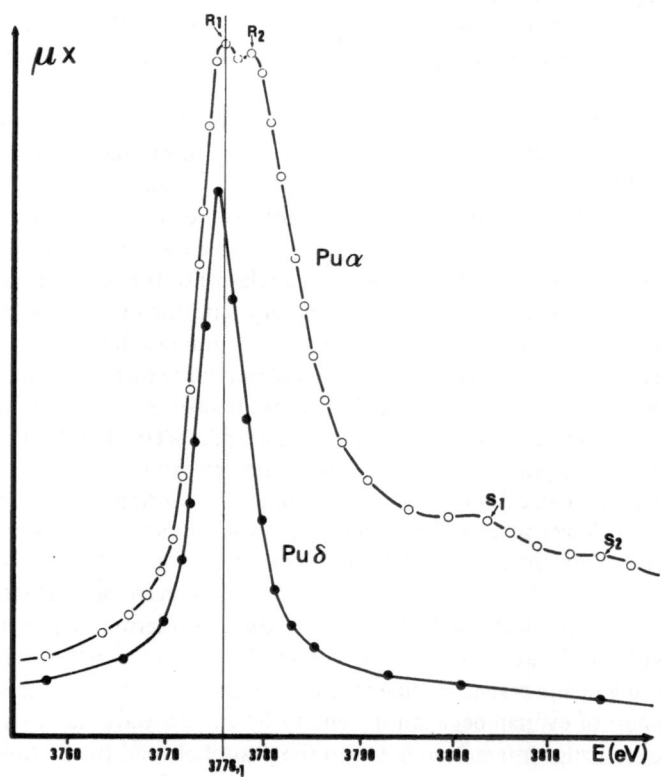

Fig. 7. M_V absorption spectra of α and δ Pu.

sorption curve is similar to that observed for the rare earths; no absorption jump is observed while a jump is visible in the spectrum of α–Pu. The absence of the absorption jump shows that the empty f states are not mixed with the conduction states. Thus, the M_V absorption line describes transitions to f states strongly localized on each ion, i.e. transitions $3d_{5/2}^6 \, 5f^n \rightarrow 3d_{5/2}^5 \, 5f^{n+1}$ in the free ion.

On the other hand, the α–Pu M_V photoabsorption is almost similar in form to a transition metal p photoabsorption. The jump denotes the presence of transitions to f states hybridized with states belonging to a continuum. This is probably also true for the S_1 and S_2 structures. One resonance line has been observed in the M_V emission spectrum in coincidence with the R_2 peak and none in the M_{IV} spectrum. It seems that in α–Pu, only a part of empty $5f$ states are localized; there are states toward which a $3d_{5/2}$ electron can be excited. All the others are $5f$–$6d$ hybridized states; these are, in particular, states toward which a $3d_{3/2}$ electron has a large probability of being transferred. Thus, the absorption curves present a jump just at the Fermi level and their inflexion point gives the position of this level.

In the δ–Pu phase, the M_V absorption line shifts toward lower energies. This shift reveals an increase in the localization of $5f$ electrons. Indeed, the electronic screening of the nuclear charge is more efficient as the number of localized electrons goes up. This increase is accompanied by a decrease in the $3d_{5/2}$ binding energy and consequently, a shift in the M_V absorption line, which is what is in fact observed. This localization produces a stabilization of f states which can decrease the relative number of empty f states.

Finally, the modification observed between α and δ phases can be interpreted by an increase in the localization of $5f$ states in δ–Pu. This is in agreement with the suggestion by *Friedel* (*36*) according to which the $5f$ electrons are appreciably more localized about each ion in the high-temperature phases than in monoclinic α–Pu. Our results confirm that the expansion in the atomic volume of the δ-phase with respect to the α-phase (see Fig. 8) is associated with an increase in the number of localized electrons forming a screen to the nuclear charge. The same modifications are likely in the intermetallic compounds where an increase in the Pu interatomic distance is stabilized, for example, by the introduction of small quantities of trivalent elements (*37*).

Lastly, it is known that the relaxation processes decrease considerably with increase in temperature. This could help to explain the difference between α and δ-plutonium. A resonance line should be expected in the M_V emission spectrum of δ–Pu. Its observation is experimentally difficult and shall be attempted in the near future.

Energy band calculations. The conduction band is characterized by a 7 sp band hybridized with a broad $6d$ band which overlaps the Fermi level. Band structure calculations have been carried out by means of the symmetrized relativistic APW method (SRAPW method) (*5*). The exchange contribution to the potential is approximated by either the Slater exchange term or by multiplying this term by $\alpha = 2/3$. The position of $5f$ states, and also their localization in energy, depends strongly on the choice of the exchange parameter, α. The radial charge density of the f character has,

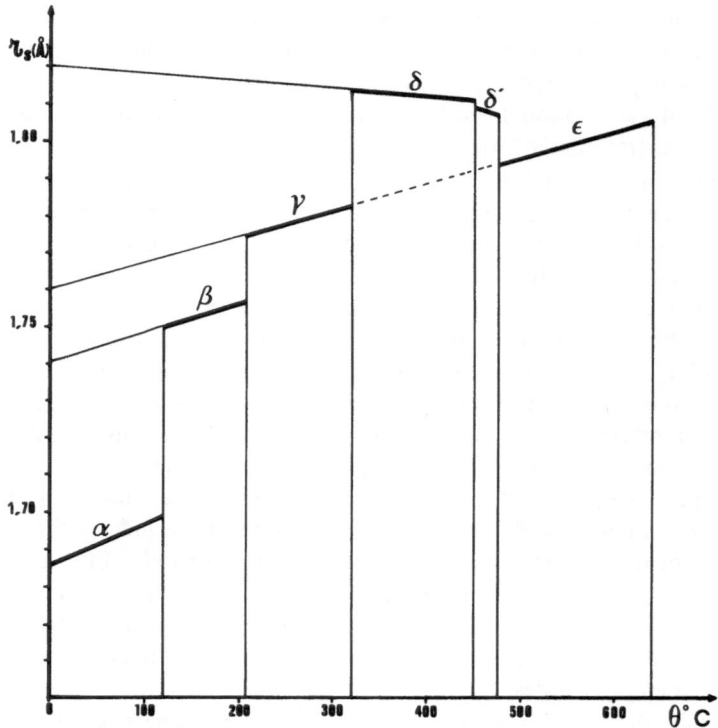

Fig. 8. Atomic radius r_s of Plutonium.

moreover, been determined at several energies throughout the band structure range (*38*). At the top of the *f* states region, the charge density is like an atomic 5 *f* level confined to a single site, but this is neither at nor below the Fermi energy. Thus, the behaviour of the *f*-character electrons depends simultaneously on the potential, i.e. the correlation and exchange terms, and on the energy of the 5 *f* states. The larger the exchange parameter and the higher the position of the 5 *f* states, the greater the localization.

From calculations for *fcc* thorium metal in the $6d^2 7s^2$ configuration with $\alpha = 2/3$, (*39*) the Th 5 *f* states are located above the Fermi level, between 0.55 and 0.6 Ryd., and are found to hybridize with the $6d$ and $7sp$ bands. On the other hand, the $6d$ band extends over about 0.7 Ryd. with a relatively large density at the Fermi level.

Energy-band structure studies have been reported on *bcc* γ-uranium metal only; the room-temperature phase has a low symmetry which adds to its complexity. The calculations have been performed for three possible configurations, $f^4 d^0 s^2, f^3 d^1 s^2$ and $f^2 d^2 s^2$, and with two values of the exchange parameter, $\alpha = 1$ and $\alpha = 2/3$ (*40*).

40

The energy of the f-states moves up with increasing $5f$ electron number and the sensitivity of the position of the $5f$ states is greater in the $\alpha = 1$ case. The "$5f$" bands are found to be hybridized with the very broad $7sp$ band and the broad $6d$ band, but for $\alpha = 1$ they appear to be very flat. For $\alpha = 2/3$, the "f" bands are wider and the width remains lower than 0.1 Ryd; then the hybridization with the d band increases. According to A. J. *Freeman*, the uranium band structure could be understood as that of a transition metal with $5f$ bands superimposed on the $d-s$ bands and hybridized with them.

The same results were obtained for Pu: calculations of the energy band structure were made for the high symmetry phase *fcc* δ–Pu in the configuration $f^5 d^1 s^2$ and the change of α from 2/3 to 1 changes the character of the $5f$ orbitals from itinerant to local ones (*41*).

Then, the conduction band structure of an actinide metal appears to be more complicated than that of a transition or rare earth metal because some $5f$ states are hybridized with the $6d$ band. According to A. J. *Freeman*, for the lighter actinides up to Pu, the degree of the overlap between the $5f$ wave functions on neighbouring atoms is large; thus the bandwidth, because of overlap and the hybridization with the $6d-7s$ bands, is noticeable. This seems to be in disagreement with the presence of R lines in the M_V and M_{IV} emission spectra. Indeed, the R lines involve $5f$ states normally empty in the unperturbed metal, and information on the localization of only $5f$ excited states is obtained by their observation and not on that of $5f$ states situated below the Fermi level.

Lastly, a localized model with a large $f-d$ hybridization has been proposed by *Jullien et al.* (*42*). In this model, two virtual bound states are present; they are hybridized $6d$ and $5f$ states, and their characteristics can explain the magnetic properties of actinides.

Energy diagram. The analysis of the M_{III} spectrum ($3p_{3/2}$) gives the distributions of the $6d - 7s$ conduction states and the position of the Fermi level. Because the width of the M_{III} level is too large for a precise determination of the conduction band shape, we have not studied the emission but only the absorption (*43, 44*). Our results show a large density of d states at Fermi level.

The M_{IV} and M_V emissions give the position of the $5f^n$ distributions, whatever the characteristics of these states – if they are localized, the hole is filled very rapidly by an Auger-type process – if not, the mobility of the hole is large; thus, at the final state of the process, the average configuration on each atom has a strong probability of being that of unperturbed metals as this is always the case for the emission band of metals.

As for the rare earths, it is possible to obtain the energy distance D between the $5f^n$ occupied distributions and the Fermi level, taking into account the energy of the M_{IV} or M_V emission and that of the inflexion point of the M_{III} absorption curve. The energy differences $(E_{M_{III}} - E_{M_{IV}})$ or $(E_{M_{III}} - E_{M_V})$ are deduced from reference (*45*). The limitations are the same as those discussed for the rare earths. The precision is not better than 1 eV.

41

In the same way we have determinated the position of resonance lines with respect to the Fermi level. If one supposes that the perturbations caused by $3p$ and $3d$ holes are the same, it appears that the $5f$ excited states involved in the R lines are situated largely beyond the Fermi level for Th and Pu. Then the resonance lines are not of an excitonic type; moreover, they are not due to a stabilization of $5f$ empty states by the inner hole because the excited states should then drop below E_F. On the contrary, in α–Pu, the R line is in coincidence with the R_2 absorption line and not with the R_1 line. However, a lowering of the position of $5f$ excited states as a consequence of the formation of the inner hole must be present and the distance between the resonance line and the Fermi level gives only a *minimum* value of the $5f^{n+1}$ states $-$ E_F distance.

An energy diagram is plotted in Fig. 9 for the three metals. A high density of $5f$ localized states is present above the Fermi energy in the thorium metal. The maximum of this distribution is at about 3 eV from E_F.

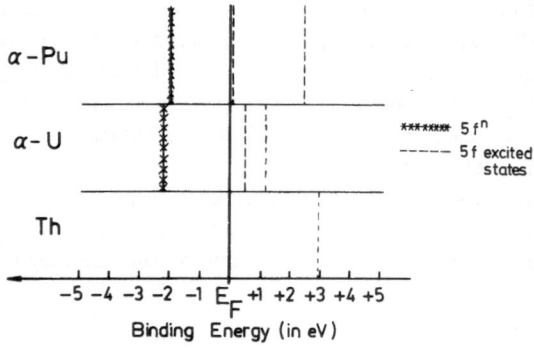

Fig. 9. Diagram of occupied and excited $5f$ states in Th, U, and Pu with respect to the Fermi level.

For α–U, the $5f$ occupied states are found at 2.2 eV below the Fermi level and other $5f$ localized states are immediately above, at 0.4 eV from the M_{IV} spectrum and 1.2 eV from the M_V spectrum.

In α–Pu, the intense atomic line, Mγ, prevents the observation of the $5f^n$ distribution from M_V spectrum (see Fig. 5). As no atomic or resonance line is present in the M_{IV} spectrum, the $5f$ occupied electron distribution is deduced more conveniently from this spectrum. Then the maximum of the $5f^n$ distribution is at about 2 eV below the Fermi level; it coincides to within ± 0.3 eV with the inflexion point of the M_{IV} absorption curve. Moreover, $5f$ excited states are present just at the Fermi level. These results confirm that the excited states toward which the $3d_{3/2}$ electrons are promoted are strongly hybridized with the $6d$ states. As to the M_V R line, it is situated at 2.5 eV beyond the Fermi level.

4.2. Compounds

We report in this paragraph the results obtained by X-ray spectroscopy for the stable oxides, ThO_2, UO_2, and PuO_2. All three compounds crystallize in the CaF_2-type structure. We have analyzed their M_{IV-V} and M_{III} absorption spectra and M_{IV-V} emission spectra of ThO_2 and UO_2.

In Fig. 10 are plotted the UO_2 M_{IV-V} emissions for different excitation voltages and the corresponding absorptions. Fig. 11 shows the comparison between the M_{IV-V} emissions of the metal and the oxide (28). The resonance lines are more intense in UO_2 than in U. As for the metal, when the excitation voltage increases, the intensity of the R lines decreases with respect to the atomic lines and to the $5f^n$ emission, but the M_V R line remains observable at a high voltage. Inversely, the $5f^n$ emissions are weaker in the oxide; they are difficult to observe and appear only when the excita-

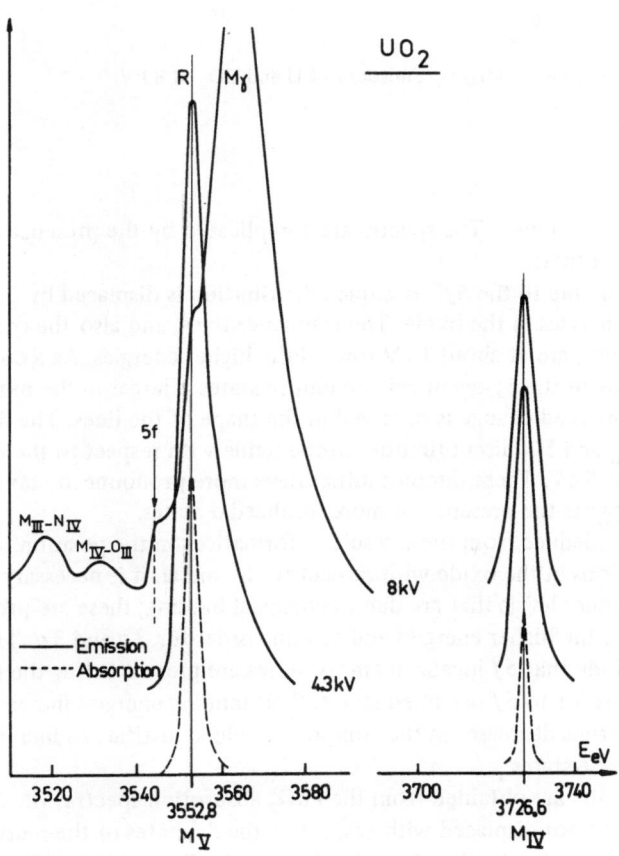

Fig. 10. M_{IV-V} UO_2 spectra: absorption and emission at 4.3 and 8 kV.

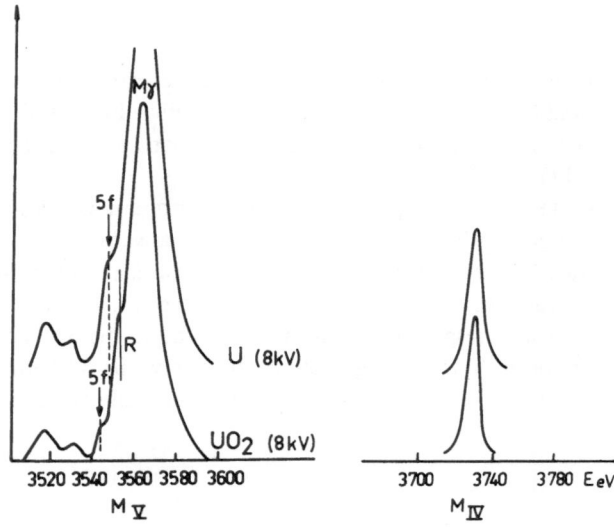

Fig. 11. Comparison between M_{IV-V} emissions of U and UO_2 at 8 kV.

tion voltage is great enough. The spectra are complicated by the presence of atomic lines, Mγ and the others.

The transition due to the $5f^n$ occupied distribution is displaced by 3.5 eV toward the lower energies in the oxide. The resonance lines, and also the corresponding absorption lines, are at about 1 eV toward the higher energies. As a consequence, the distance between the $5f$-occupied and empty states is larger in the oxide; it is about 7 eV. No marked change is observed in the shape of the lines. The displacements of the M_{III} and M_{II} discontinuities in the oxide with respect to the metal are approximately of 5 eV. These discontinuities have more pronounced maxima in the oxide, which suggests the presence of more localized d states.

To be able to deduce from these results information on the position of $5f$ and $6d-7s$ distributions in the oxide with respect to the metal, it is necessary to know the shifts in the inner levels that are due to chemical binding; these are probably several eV toward the higher energies and of same order for $3p$ and $3d$. Thus, it is possible to conclude that $5f$ localized empty states are present below the conduction band in the oxides. As to $5f$ occupied states, their binding energies increase. The number of f electrons decreases in the compound, which justifies an increase in the localization of these states.

The same results are obtained from the PuO_2 absorption spectra: the $5f$ empty states are pratically not displaced with respect to the $5f$ states of the metal, whereas the bottom of the conduction band is + 5 eV above the Fermi level of Pu metal. Thus, the $5f$ empty states are situated below the conduction band in PuO_2.

For ThO_2, we have observed no displacement of M_{IV-V} R and absorption lines. As for the M_{III} and M_{II} discontinuities, they are displaced by about 1 eV only toward the higher energies. Thus, the $5f$ empty states are above the bottom of the conduction band in this compound.

A particular point must be emphasized: we have observed toward the lower energies of the ThO_2 R lines, a weak emission which seems to correspond to the transition from the f states to the $3d$ level (46). From optical spectra, one has determined that the energy of the $5f$ states increases with the degree of ionization and becomes higher than the energy of the $6d$ states for the Th^{3+} ion whose configuration is $[Rn]\,5f^1$. As a consequence, it is possible to suggest that in ThO_2 a few f orbitals are mixed with the oxygen $2p$ orbitals. Thus, the bonding is not purely ionic in this compound – a contribution of the covalency must be involved, an electron charge transfer taking place to the thorium ion from the more electronegative oxygen atoms.

In summary, because the inner shell energy levels shift in the direction of higher binding energy by an unknown quantity, only qualitative information is obtained on the relative position of the electron distributions between metal and compound. However, provided the shift is approximately the same for $3p$ and $3d$, it is possible to deduce from our results an energy diagram for each oxide (cf. Fig. 12, where the energy zero is the bottom of the conduction band).

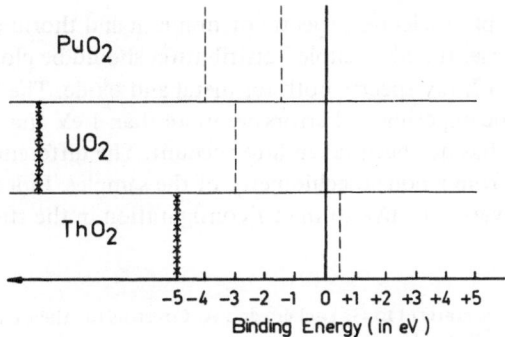

Fig. 12. Diagram of occupied and excited $5f$ states in ThO_2, UO_2, and PuO_2 with respect to the bottom of the conduction band.

On the other hand, the degree of localization of the $5f$ electrons increases in the oxide, especially in UO_2. Indeed, the localization increases with the internuclear distance because the overlap of neighbouring $5f$ orbitals decreases. However, an absorption jump is always observed in the absorption spectra of oxides; the $5f$ localization is therefore weaker in the light actinides than in the rare earths, except for δ–Pu.

4.3. Comparison with photoelectron spectroscopy

X-ray spectroscopy and photoelectron spectroscopy are two different methods for the analysis of electron distributions in solids. Each of them brings complementary information and possesses distinctive characteristics. Three points can be emphasized.

Firstly, the electron spectroscopy provides information on the occupied states distributions only, whereas X-ray spectroscopy involves both filled and empty states.

Secondly, photoelectron spectroscopy gives the distribution of all mixed states and the spectrum is disturbed by the photoionization probabilities of the various symmetries; as a consequence, it is sometimes difficult to interpret the results. On the other hand, the X-ray spectra involve the distributions of known symmetries, i.e. those states from which the transitions toward the inner hole are possible considering the transition probabilities.

Lastly, in the two methods, the spectra provide a description of the final state. When an outer incomplete localized sub-shell is present, as is the case in the rare earths, the kinetic energy of photoelectrons reflect the final state configuration with $(n - 1)$ electrons and possesses a multiplet structure (47). Then, in the rare earth $4f$ photoelectron spectra, the final state is $4f^{n-1}$; the observed demultiplication is well interpreted by this configuration (48). We have seen that in the process of the M_{IV-V} emission and absorption, the final state has either the $4f^n$ or $4f^{n+1}$ configuration. Indeed, the differences observed between the various spectroscopic results are compatible with this interpretation.

For actinides, only the photoelectron spectra of uranium and thorium have been analyzed (49, 50). From these, the $5f$ occupied distribution should be closer to the Fermi level than found from X-ray spectra both for metal and oxide. The difference is 2 eV for the metal, but the experimental errors are more than 1 eV and the perturbation due to inner hole has not been taken into account. The difference is greater in the oxide; this can arise from a non-stoichiometry of the samples. Indeed, the XPS and X-ray spectra are very sensitive to the $5f$ configuration in the studied solid.

Acknowledgements: The writer is grateful to G. Lackérè and A. Courtois for their experimental work on the X-ray spectroscopy of Th, U and Pu.

References

1. *Jayaraman, A.:* Phys. Rev. *137A* (1965) 179.
2. *Jørgensen, C. K.:* Structure and Bonding *22* (1975) 49.
3. *Nellis, W. J., Brodsky, M. B.:* The Actinides, Vol. II, Chap. 6. New York: Academic Press. 1974.
4. *Frazer, B. C., Shirane G., Cox, D. E., Olsen, C. E.:* Phys. Rev. *140A* (1965) 1448.
5. *Freeman, A. J., Koelling, D. D.:* The Actinides, Vol. I, Chap. 2. New York: Academic Press 1974.
6. *Yariv, A.:* Quantum Electronics. New York: John Wiley and Sons, Inc. 1967.
7. *Bonnelle, C.:* Ann. Phys. *1* (1966) 439.
8. *Anderson, P. W.:* Phys. Rev. Letters 18 (1967) 1049.
9. *Bergersen, B., Brouers, F., Longe, P.:* J. Phys. *F1* (1971) 945.
10. *Arakawa, E. T., Williams, M. W.:* Phys. Rev. Letters *36* (1976) 333.
11. *Bonnelle, C., Karnatak, R. C.:* J. Physique *32* (1971) C4-230.
12. *La Villa, R. E.:* Phys. Rev. *A9* (1974) 1801.
13. *Bonnelle, C.:* Physica Fennica *9S1* (1974) 92.
14. *Dufour G., Bonnelle, C.:* J. de Phys. Lettres *35* (1974) L-255.
15. *Sugar, J.:* Phys. Rev. *A6* (1972) 1764.
16. *Dehmer, J. L., Starace, A. F., Fano, U., Sugar J., Cooper, J. W.:* Phys. Rev. Letters *26* (1971) 1521.
17. *Karnatak, R. C.:* These de Doctorat d'Etat, Paris, 1971.
18. *Mariot, J.M., Karnatak, R. C.:* J. Phys. *F4* (1974) L223.
19. *Bonnelle, C., Karnatak, R. C.:* C.R. (Paris) *268* (1969) 494.
20. *Bonnelle, C., Karnatak, R. C., Sugar, J.:* Phys. Rev. *A9* (1974) 1920.
21. *Bonnelle, C., Karnatak, R. C., Spector, N.:* (in press).
22. *Mariot, J.-M., Karnatak, R. C.:* Solid State Comm., *16* (1975) 611.
23. *Mariot J.-M., Karnatak, R. C.; Bonnelle, C.:* J. Phys. Chem. Solids *35* (1974) 657.
24. *Bonnelle, C., Karnatak, R. C., Jørgensen, C.K.:* Chem. Phys. Letters *14* (1972) 145.
25. *Muller, N., Siegmann, H. C.:* Phys. Letters *24A* (1967) 733.
26. *Bonnelle, C., Vergand, F.:* J. Phys. Chem. Solids *36* (1975) 575.
27. *Courtois, A.:* These CNAM, Paris, 1975.
28. *Bonnelle, C., Lachere, G.:* J. de Phys. *35* (1974) 295.
29. *Bonnelle, C., Lachere, G.:* under preparation.
30. *Bonnelle, C., Lachere, G.:* 5th International Conference on Plutonium and Other Actinides, North-Holland Amsterdam 1976.
31. *Bonnelle, C., Belin, E.:* under preparation.
32. *La Villa, R. E.:* J. Chem. Phys. *57* (1972) 899.
33. *Nordgren, J., Werme, L. O., Ågren, H., Nordling, C., Siegbahn, K.:* J. Phys. *B8* (1975) *L18*.
34. *Larson, D. T., Adams, R. O.:* Surface Science *47* (1975) 413.
35. *Bonnelle, C., Courtois, A., Calais, D.:* 5th International Conference on Plutonium and Other Actinides, North-Holland, Amsterdam 1976.
36. *Friedel, J.:* Rapport CEA n° *766* (1958) Paris.
37. *Calais, D., Zanghi, J.P.,* J of Nuclear Materials *39* (1971) 350.
38. *Tucker, T. C., Roberts, L. D., Nestor, C. W., Carlson, T. A., Malik, F. B.:* Phys. Rev. *174* (1968) 118.
39. *Koelling, D.D., Freeman, A.J.:* Solid State Comm. 9 (1971) 1369.
40. *Koelling, D.D., Freeman, A.J.:* Phys. Rev. *B 7* (1973) 4454.
41. *Freeman, A.J., Koelling, D.D.:* J. Physique *33* (1972) C3-57.
42. *Jullien, R., Galleani d'Agliano, E., Coqblin, B.:* Phys. Rev. B6 (1972) 2139.
43. *Cauchois, Y., Bonnelle, C., Manescu, I.:* C.R. Paris *267B* (1968) 817.
44. *Cauchois, Y., Bonnelle, C., de Bersuder, L.:* C.R. Paris *256* (1963) *112* and *257* (1963) 2980.

C. Bonnelle

45. *Cauchois, Y.:* J. Physique et Radium *13* (1952) 113.
46. *Bonnelle, C., Lachere, G.:* in press.
47. *Spector, N., Bonnelle, C., Dufour, G., Jørgensen, C.K.; Berthou, H.:* Chem. Phys. Letters *41* (1976) 199.
48. *Cox, P.A., Baer Y., Jørgensen, C.K.:* Chem. Phys. Letters *22* (1973) 433.
49. *Verbist, J., Riga, J., Pireaux, J.J., Caudano, R.:* J. of Electron Spec. *5* (1974) 193.
50. *Veal, B.W.:* The Actinides, Vol. II, Chap. 3. New York: Academic Press 1974.

Application of the Functional Approach to Bond Variations under Pressure

V. Gutmann and H. Mayer

Institut für Anorganische Chemie und Institut für Mineralogie, Kristallographie und Strukturchemie der Technischen Universität Wien, Austria

Table of Contents

1. Introduction

It is well known that many materials, whether they have originally ionic, non-ionic, or molecular lattice structures, are transformed into the metallic state by the application of sufficiently high pressure, and indeed this can be expected to be true of all materials. Even quite modest increases in pressure can affect interatomic distances, spectral transitions, formal oxidation states, and many other phenomenological parameters, e.g. can increase the coordination number. Various attempts have been made in an effort to establish relationships between pressure and these phenomenological parameters but none of them accounts satisfactorily for all of the observed features. This is almost certainly because of the absence, up to now, of a model which is capable of interpreting the facts without concerning itself with too detailed an interpretation of the binding forces. However, it will be shown here, after a brief survey of the present situation, that the "functional approach" seems to successfully provide such a model based as it is on an electron-pair donor-acceptor model of molecular interactions.

2. Present Interpretations

2.1. The Pressure-Coordination Rule

The most useful rule in describing the effect of pressure on solids is the so-called "Pressure-Coordination Rule" (1, 2) according to which the coordination number is increased with pressure. In Table 1 and 2 examples are listed for various crystal structure transformations which follow this qualitative rule at different pressures and temperatures. An exception to this rule is known, however, for ytterbium (3); the cubic face-centered modification (coordination number = 12) of the metal is transformed at 40 kbar into a cubic space-centered structure (coordination number = 8).

2.2. The Pressure-Homology Rule

A special case of the pressure-coordination rule is the pressure-homology rule (1, 2). This can be applied when comparing crystal structures within one homologous series of compounds, and states that the crystal structure of the heaviest homologue is obtained for the lower homologues by an increase in pressure. For example the crystal

Table 1. Phase transitions under pressure. Coordination numbers in [ ]

Normal Pressure Phase	High Pressure Phase	Conditions for Transformation		References
		Pressure [kbar]	Temp. [°C]	
1. Transformation [3] → [4]				
Graphite	Diamond	160	3500	(4)
BN (Graphite type)	BN (Diamond-type)	60	1350	(5)
2. Transformations [4] → [6]				
Diamond	C (metallic)	650	1100	(6)
Quartz ⎫ Coesite ⎬ SiO$_2$	Stishovite (Rutile type)	100	1300	(7, 8)
GaAs (Zinc blende-type)	GaAs (NaCl-type)	≳ 250	20	(9, 10, 11)
ZnO (Wurtzite type)	ZnO (NaCl-type)	110	20	(9, 10, 11)
CdTe (Zinc blende-type)	CdTe (NaCl-type)	≈ 10	20	(9, 10, 11)
3. Transformation [5] → [6]				
Andalusite	Cyanite	7	200	(9)
4. Transformations [6] → [8]				
RbCl ⎫ RbBr ⎬ (NaCl-type) RbI ⎭	CsCl-type	5	20	(9, 10)
KCl ⎫ KBr ⎬ (NaCl-type) KI ⎭	CsCl-type	≈ 20	20	(9, 10)
MnF$_2$ (Rutile-type)	CaF$_2$-type	14	500	(9)
5. Transformation [6] → [9]				
Calcite	Aragonite	10	400	(12)
6. Transformations [8] → [12]				
Cs ⎫ cubic Ba ⎬ body-	Cs ⎫ cubic Ba ⎬ face-centered	59	20	(9, 10)
α-Fe ⎭ centered	⎰ Fe (cub.) ⎱ Fe (hex.)	133	20	(9, 10)

structure of cesium chloride (coordination number = 8) stable at room temperature is obtained for NaCl, KCl and RbCl under pressure. The required pressure is greater the lighter the constituents. According to Table 2 the pressures required increase in the series RbCl < KCl < NaCl as the cation radius is decreased. The same trends are seen in the more covalent series InSb < GaSb < AlSb and ZnTe < ZnSe < ZnS (Table 3).

For the alkali metal halides the influence of the anion is practically non-existent as can be seen from the comparison of RbCl, RbBr and RbI (Table 3). In the more covalent structures both atoms appear to be effected by increase in pressure, as changes in the band structure lead to an increase in conductivity.

Table 2: Changes in coordination number under pressure for chemical reactions
(*13, 14*)

[6] [6] [4][4] (MgAl) [AlSiO$_3$] Al-Enstatite (mmm, chain)	$\geqq$ 20 kbar $\geqq$ 1200 °C	[6][4] [6] Mg[SiO$_3$] + Al$_2$O$_3$ Enstatite Corundum (mmm, chain) (3 m)
[6] [4] [8] [4][4] Mg$_2$[SiO$_4$] + Ca$_2$[AlSi$_2$O$_8$] Forsterite Anorthite (mmm, insula) (1)	$\geqq$ 20 kbar $\approx$ 1000 °C	[8] [8][6] [4] Mg$_2$CaAl$_2$[Si$_3$O$_{12}$] Garnet (m3m)
[4][4] 2 LiAlSiO$_4$ α-Eucryptite (62, insula)	55 kbar (+ 5% LiF) 1000 °C	[6] [6] [6] [6] Li AlSi$_2$O$_6$ + Li AlO$_2$ Spodumene (2/m, chain) (trigonal)
[4] [4] Na[AlSi$_3$O$_8$] + Na[AlSiO$_4$] Albite Nepheline (1) (6)	$\approx$ 18 kbar 1000 °C	[6] NaAl[Si$_2$O$_6$] Jadeite (2/m, chain)
[6] [4] 2 Mg SiO$_3$ Klinoenstatite (2/m, chain)	115 kbar 650 °C	[6] [4] [6] Mg$_2$SiO$_4$ + SiO$_2$ Forsterite Stishovite (mmm, insula) (4/m)

Table 3. Coordination Changes and Pressure-Homology Relationships[15]

Formula	Type	Coord. Number	Pressure [kbar]	Temperature [°C]	Coord. Number	High- Pressure Type
NaCl	NaCl	6	100	20	8	CsCl
KCl	NaCl	6	20	20	8	CsCl
RbCl	NaCl	6	5	20	8	CsCl
RbBr	NaCl	6	5	20	8	CsCl
RbI	NaCl	6	5	20	8	CsCl
AlSb	D	4	125	25	6	Sn$_w$
GaSb	D	4	90	25	6	Sn$_w$
InSb	D	4	22	25	6	Sn$_w$
ZnS	D	4	240	20	6	NaCl
ZnSe	D	4	165	20	6	NaCl
ZnTe	D	4	140	20	6	NaCl

Table 4. Transformations under Pressure within the homologous series C–Si–Ge–Sn

Normal Pressure Phase			Pressure [kbar]	Temperature [K]	High Pressure Phase		
C	(Diamond)	[4]	$\geqslant 650$	≈ 1100	C	(metallic)	[6]
Si	(D-type)	[4]	≈ 200	≈ 300	Si	(Sn_w-type)	[6]
Ge	(D-type)	[4]	≈ 120	≈ 300	Ge	(Sn_w-type)	[6]
Sn_g	(D-type)	[4]	≈ 1	$\gtrsim 286$	Sn_w		

In molecular lattices, chain lattices and layer structures the coordination numbers are found to be increased by applying sufficient pressure, and at very high pressure a highly coordinated metallic packing is obtained. For example phosphorus is converted into a structure in which its homologous element, antimony, is stable at ordinary pressure (16).

Within the homologous series C–Si–Ge–Sn all of the elements are stable (or metastable) in the diamond structure (coordination number = 4). Pressure converts these non-conductors or semi-conductors into metallic modifications with coordination number of 6 (which is the stable modification of tin at room temperature and atmospheric pressure (6, 17, 18) (Table 4).

2.3. The Pressure-Distance Paradox

The basic effect of pressure is of course to increase the density of the material. Hence decrease in bond-lengths should be expected. The "Pressure-distance Paradox" expresses the fact that this may not be true of *all* interatomic distances, as the increase in pressure may in fact cause a *lengthening* of the shortest interatomic (or bond) lengths in the crystal. For example, the transformation of coesite (SiO_2) with coordination number of four into stishovite (SiO_2) with a coordination number of six is accomplished at $\geqslant 100$ kbar at $\geqslant 1200\,°C$ (7, 8). As expected the density increases (from 2.93 to 4.28 g cm^{-3}, i.e. by 46%) but the Si–O bond lengths are *increased* from 161.3 pm in coesite to 177.8 pm in stishovite, an increase of 10.2%. Another well-known example of this paradox is the transition of graphite into diamond (4); the C–C distances in diamond are 8.7% *greater* than those within the graphite layers, although the density is of course considerably increased in going from graphite to diamond due to the far greater *shortening* of the interlayer distances initially present in graphite.

2.4. Pressure Induced Electronic Transitions

In order to account for this effect the electronic transitions provoked by increase in pressure must be considered, and it is interesting to note that changes in electronic energy levels are found to occur in charge transfer complexes as well as in transition metal complexes.

Jørgensen (20) has analyzed a number of spectra of hexacoordinated transition metal ions in relation to pressure. In K_2OsBr_6 the peaks at 22.200 cm^{-1} and at 16.750 cm^{-1} are shifted distinctly to lower energy with increasing pressure; at the same time the splitting increases drastically. Spectroscopic changes occur as a result of the pressure induced transformation of tetrahedral cobalt(II)-complexes in solution into octahedral arrangements (21, 22), and indeed are used as evidence for this transformation.

In many compounds of iron(III), the application of pressure causes a change in the spectroscopic oxidation state of the iron(II), i.e. an apparent reduction is observed. The mechanism of this reduction is thought to involve the transfer of an electron thermally from a ligand non-bonding level to the metal d_π-orbitals (23). A good example is afforded by the use of Mössbauer spectra for following the effect of pressure on Prussian Blue: the iron normally exists in two distinct oxidation states with the possibility of electron transfer between them (23). At 145 kbar pressure the typical high-spin iron(III) spectrum is changed to that of the high-spin iron(II) state. At 23 °C and 145 kbar pressure 85% of the iron(III) has been reduced, and the electron transfer across the cyanide bridge seems to depend but little on temperature in contrast to the other processes involved (23).

All of these phenomena support the view that pressure acts by increasing the electron donor properties of parts of the systems. In fact in many ways the application of pressure can be said to fulfil the role of an electron donor, but no unified approach has been presented from this point of view so far.

3. The Functional Approach

3.1. Basic Considerations

Based on well-known facts and ideas, the so-called functional approach emphasizes the actual electronic functions by each of the interacting molecules (7), atoms or ions (24). These cannot be understood by the exclusive consideration of the structures of the isolated entities, as their functions are highly influenced by the mode of the mutual interactions.

These may be considered as donor-acceptor interactions in the broadest sense (25); a given species may function either as donor or as acceptor for electrons. Any such interaction leads to characteristic electronic rearrangements with the reacting entities (25). It is a fact that a heteronuclear bond $\overset{\delta+}{A}-\overset{\delta-}{B}$ is attacked by an electron donor at the area of low electron density, e.g. at A^+, while an electron acceptor will interact at the area of high electron density, e.g. at B^-. In the first case the electron density at A is increased and in the second case, that at B is decreased. By each of these changes an electron shift is induced from A to B which leads to polarization of the existing bond A–B which manifests itself as an increase in bond length: (25)

$$\text{Donor} \text{———} \underset{A}{\overset{\delta+}{}} \xrightarrow{} \underset{B}{\overset{\delta-}{}}$$

$$\underset{A}{\overset{\delta+}{}} \xrightarrow{} \underset{B}{\overset{\delta-}{}} \text{———→ Acceptor}$$

Induced
Increase in
Bond length

(The slightly bent arrow indicates the direction of the induced electron shift)

The electron donor may act either as a Lewis base or as a reducing agent. For example the B–F distances in BF_3 are increased by complex formation with dimethylsulfoxide from 130 pm to 136 pm.

Lewis   Coordination   Increase
Base   in B–F–
                                 bond length

Likewise the Fe–O bond length in their hydrated iron(III)-ion is increasing on reduction to Fe(II). By analogy, the electron acceptor may act either as a Lewis acid or as an oxidizing agent.

These rules may be applied also to hydrogen bonding and even to van der Waals type interactions.

Prout and *Wright* (26) pointed out that the bond polarization induced by coordination is greater the shorter the coordinate bond (Table 5).

Table 5. Bond lengths in BF_3 and BF_3 adducts [pm]

Bond length	BF_3	$CH_3CN \rightarrow BF_3$	$H_3N \rightarrow BF_3$	$Me_3N \rightarrow BF_3$
$-N-B-$	–	163	160	158
$-B-F$	130	133	138	139

Based on experimental evidence in various systems the following rule has been formulated (25): *The stronger the intermolecular interaction, the greater is the extent of bond polarization of the intramolecular bonds adjacent to the new bond:*

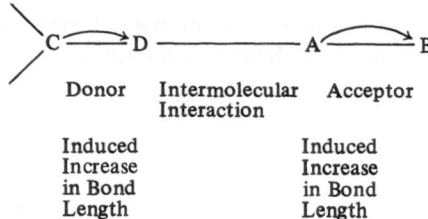

Donor	Intermolecular Interaction	Acceptor
Induced Increase in Bond Length		Induced Increase in Bond Length

This rule is also obeyed by molecules with homonuclear bonds (25) (except by molecular oxygen). For example the intramolecular I–I-distance in gaseous iodine is 267.6 pm (27). However, in the solid state the interaction between the molecules as characterized by the intermolecular distance of 354 pm induces an increase in intramolecular bond length to 271.5 pm (28).

By these molecular interactions the coordination numbers are increased. From this point of view the following rule may be formulated: *Increase in coordination number leads to lengthening of the bonds which already exist at the coordination center.*

The essential points concerning the wide applicability of these rules are as follows (29):

1. Emphasis is laid on observable changes in bond parameters as the result of the electronic rearrangements.
2. The changes are independent of the interpretation of the bonding forces. Changes in covalent, ionic, metallic, hydrogen bonds, or van der Waals type interactions are subject to the same rule to various degrees.

3. The rules are in no way contradicted by ideas derived from quantum chemistry and can be applied to very weak (physical) as well as to strong (chemical) interactions.

The importance of the application of these rules has been demonstrated for all of the aspects of the chemistry in non-aqueous solutions, as well as for catalytic and adsorption phenomena, Based on Nolls ideas (30) the application to silicate structures has recently been demonstrated (25).

We shall now attempt to apply these rules to the electronic rearrangements due to the action of pressure on crystal structures by considering the function of pressure as equivalent that of an electron donor.

3.2. Application to Pressure Effects in Molecular and Layer Lattices

We may start from the rule which relates the increase in the coordination number to the increase in adjacent bond lengths. In a molecular lattice a distinction between intermolecular and intramolecular bonds is made. In a layer lattice, such as graphite an analogous distinction may be made between interlayer bonds and intralayer bonds.

In both of these cases the concept of induced bond polarization may be applied as follows: Pressure is considered as equivalent to the action of an electron donor which increases the electron densities in the first place along the intermolecular and interlayer bonds respectively. The latter are considerably shortened and hence a geometrical rearrangement with increase in the coordination number may also take place. Both the shortening in intermolecular bond length and the increase in the coordination number induce lengthening of the intramolecular or intralayer bond lengths respectively. The observed increase in density is due to the fact, that the decrease in intermolecular bond lengths is greater than the induced increase in intramolecular bond lengths.

From this point of view the pressure-distance paradox is a logical consequence of the functional approach!

In crystalline iodine pressure leads to changes in macroscopic properties which are characterized by an increase in metallic properties. This suggests an increase in the coordination number with simultaneous decrease in intermolecular and an increase in intramolecular bond length finally to that extent that both become equally long. The decrease in intermolecular bond length is expected to be greater than the increase in intramolecular bond length, as it is observed for the formation of the triiodide ion from iodine and iodide ion. In crystalline iodine the intermolecular I–I-distance is 354 pm and the intramolecular I–I-distance 272 pm (28). The formation of the I_3^--ion may be considered as due to the replacement of an I_2-molecule by an iodide ion: $I^- \rightarrow I–I$. Although the ionic radius of the iodide ion is considerably greater than the covalent radius of the iodine atom in the molecule, the "intermolecular I–I distance" is decreased to a greater extent than the intramolecular I–I distance is increased, both being 293 pm in the triiodide ion (31).

Structural data are available for the high pressure transformation of graphite into diamond. In the graphite structure the intralayer C—C distances may be treated like intramolecular bonds and those between the layers as intermolecular "bonds". The intermolecular distances in graphite of 335 pm are drastically decreased to 154.4 pm when transformation is taking place into the diamond structure, while the intramolecular bond distances are lengthened to 142 to 154.4 pm and at the same time the coordination number is increased from 3 to 4. The intramolecular gain in bond length is by far less pronounced than the decrease in intermolecular bond length and hence the density is increased from 2.22 to 3.514 g/cm³ (*19*).

The situation is analogous for the transformation of the hexagonal B-12 type structure of BN at ⩾ 60 kbar and 1300 °C into the cubic B-3 (zinc blende) structure (borazon). The interlayer distances are drastically decreased from 334 pm to 157 pm, while the intra-layer B—N distances are less dramatically increased from 145 pm to 157 pm. The net-effect may account for the increase in density from 2.30 to 3.45 g/cm³ (*19*).

In white phosphorus the intramolecular P—P-distances are 221 pm and the intermolecular distances considerably longer. At 12 kbar black phosphorus is produced, the intramolecular distances being increased to 222, 224 and 331 pm although no change in coordination number is involved. The interlayer distances in black phosphorus are 359 pm. When this is transformed into the As-type structure the interlayer distance is reduced to 327 pm with the coordination number unchanged. More remarkable changes occur when the coordination number is increased from 3 to 6 under a pressure of 125 kbar: all of the P—P-internuclear distances are found to be 238 pm, and this is smaller than the interlayer bond lengths and greater than the intralayer bond lengths in black phosphorus (*16, 32*).

3.3 Application to Pressure Effects in Coordination Lattices

In a coordination lattice the effect of pressure will lead to increasing bond distances (when the coordination number is increased. This may be considered as the result of the shortening of the homonuclear distances within the crystal lattice which are usually not considered as chemical bonds. If these homonuclear distances are treated like intermolecular bonds, the same approach as presented for a molecular lattice may be applied.

Both zinc sulfide and zinc selenide are converted from the zinc blende type (coordination number = 4) into the sodium chloride structure (coordination number = 6) at 117 kbar and 100 kbar respectively. In zinc sulfide the Zn—S distance is increased from 233.9 pm to 250 pm while the Zn—Zn distance is decreased from 382 to 353 pm. In zinc selenide the Zn—Se distance is increased from 246 pm to 254 pm while the Zn—Zn distance is considerably decreased, namely from 401 to 359 pm. Since the Zn—Zn distances are more strongly reduced, than the Zn—Se-bonds are lengthened, the volume is also decreased (*33*).

Both in the wurtzite and zinc blende structures (coordination number = 4) of silver iodide, stable at atmospheric pressure, the Ag—I-distances are 278 pm and 280 pm respectively. At 4 kbar the NaCl-type structure (coordination number = 6) becomes stable, in which the Ag—I-distances are remarkably greater, namely 303 pm. Increase in pressure to 100 kbar contracts all of these distances to 283 pm with a decrease in molar volume from 33.8 to 27.6 cm^3/mol. Thus the transformation with increase in coordination number from the low-pressure modification into the high pressure modification involves an increase in Ag—I-distances by 8.4%, but further increase in pressure does not produce another geometrical rearrangement and hence all of the equidistant bonds within the crystal lattice are shortened (34).

At ordinary pressure silicon displays the coordination number of four towards oxygen in most of the silicate structures. The transformation of SiO$_2$ into an octahedral arrangement is another example for the functional interpretation of the pressure-distance paradox. Coesite (coordination number = 4) is converted into stishovite (rutile-type, coordination number = 6) at $\geqslant 100$ kbar and $\geqslant 1200$ °C (7, 8). By this process both of the homonuclear distances are decreased, namely the Si—Si distances ("intermolecular") from 301 pm to 267 pm, which is close to that in oxy-gen-free K$_4$Si$_4$ (Si—Si distance = 246 pm (35) as well as the O—O-distances ("inter-molecular") from 263 pm to 251 pm. On the other hand the Si—O bond lengths are increased by 10.2% from 161.3 pm to 177.8 pm (19). The latter is overcompensated by the shortening of the Si—Si- and O—O-distances and hence the density is increased by 46.1% from 2.93 to 4.28 g/cm^3.

On the other hand the Si—O-distance in the quartz-coesite transformation (36), is only slightly increased from 160.7 to 161.5 pm, when the coordination number of four remains unchanged; the density is increased to a moderate extent, namely from 2.65 to 2.93 g/cm^3.

In magnesium silicates under normal pressure the coordination number is 6 and the Mg—O-distance 210 pm. Pyrope,

$$\overset{[8]}{Mg_3}(Al\overset{[6]}{Mg_{0.5}Si_{0.5}})(SiO_4)_3$$

is synthesized at 30 kbar. In this structure Mg has the coordination number of 8 to-wards oxygen and the Mg—O-distance is 225 pm (37).

It should be noted that apart from the high pressure modifications silicates with hexacoordinated silicon atoms are stable at normal pressure (Table 6).

The octahedral arrangement in pyridinium tris-(o-phenylendioxy)-siliconate $[C_5H_5NH]_2[C_6H_4O_2)_3Si]$ has been confirmed by the crystal structure determination. The mineral thaumasite belongs also to this group, as well as various silicon phosphates, which can be synthesized at atmospheric pressure. In stishovite and in $K(\overset{[6]}{Al_{0.25}}Si_{0.75})_4O_8$ (hollandite structure) (46) octahedra are connected by edges and corners to give the three-dimensional structure, in that way while in silicon phos-phates and in thaumasite they are separated by anions.

Table 6. Mean internuclear distances for compounds with [SiO$_6$]-octahedra

	Si[6]–O	O–O	References
Si$_5$O[PO$_4$]$_6$	1,76$_3$	2,49$_7$	(38)
SiP$_2$O$_7$, A(I)	1,75	2,48	(39)
SiP$_2$O$_7$, A(III)	1,80	2,49$_3$	(40)
SiP$_2$O$_7$, A(IV)	1,76$_3$	2,54	(41)
Ca$_3$[Si(OH)$_6$](SO$_4$)(CO$_3$) · 12 H$_2$O Thaumasite	1,78$_3$	2,52	(42)
[C$_5$H$_5$NH]$_2$[C$_6$H$_4$O$_2$)$_3$Si], Pyridinium-tris-(o-phenylendioxy)-siliconate	1,78$_4$	2,52$_7$	(43)
SiO$_2$, Stishovite	1,76$_8$	2,49$_8$	(44)
SiO$_2$, Stishovite	1,77$_8$	2,51$_3$	(45)

According to the results of the crystal structure determinations the [SiO$_6$]-octahedra appear slightly distorted. The mean value for the Si–O distance is 177 pm and for the O–O-distance 251 pm. The respective values for tetracoordinated silicon are 162 pm and 264 pm respectively. The molar volume for the hexacoordinated species is by 13% smaller than for the tetracoordinated one (47).

For the hexacoordinated species which are stable at atmospheric pressure, the electronegativities of the A-atoms in the Si–O–A sequences have been suggested to be responsible (47, 48). The greater the electron withdrawal from the oxygen atom by A (the stronger its electron acceptor properties) the more will the Si–O bond be polarized and the greater their stability (Table 7). Hence the pressure required to stabilize these compounds increases with decreasing electron acceptor properties of A.

Table 7. Relationship between electronegativites x_A of A-atoms and pressure required for the synthesis of compounds with hexacoordinated silicon

Compound	Si–O–A	x_A	pressure [kbar]
[C$_5$H$_5$NH]$_2$[(C$_6$H$_4$O$_2$)$_3$Si]	Si–O–C	2,5	10^{-3}
Ca$_3$[Si(OH)$_6$](SO$_4$)/(CO$_3$) · 12 H$_2$O	Si–O–H	2,1	10^{-2}
Si$_3$(Si$_2$O)[PO$_4$]$_6$ SiP$_2$O$_7$ AI, III, IV	Si–O–P	2,06	10^{-2}
Stishovite, SiO$_2$	Si–O–Si	1,74	90
[6][4] Mg$_3$(AlSi$_{0.5}$Mg$_{0.5}$)[SiO$_4$]$_3$	Si–O–(Mg,Al)	1,28[a]	100
[6] K(Al$_{0.25}$Si$_{0.75}$)$_4$O$_8$ (p)	Si–O–(K,Al)	1,02[a]	120

[a] arithmetic means values x_A of the respective cations

This shows in which way induced effects can be the result of the contributions of both chemical interactions and the action of pressure.

It has been mentioned that NaCl is converted from the six-coordinate structure into the 8-coordinate cesium chloride type by pressure. Indeed, the Na—Cl distances are greater in the latter than in the former. Precise structure data are available for the NaCl-type and CsCl-type structures of rubidium chloride, the latter being formed by a pressure of 11 kbar at 25 °C. In the 6 coordinate structures the Rb—Cl-distance is 329 and the Rb—Rb-distance 465 pm while in the 8 coordinate structure the former is greater (371 pm) and the latter smaller (429 pm) (*49*).

3.4. Metallic Structures

Metallic structures of coordination number = 12 are incapable of increasing the coordination number. Application of pressure results in lattice contraction with simultaneous delocalization of electrons. For example in the cubic face centered cerium metal the internuclear distances of 362 pm are contracted to 340 pm under pressure at 20 kbar (*50*).

3.5. Application to Effects on Crystal Surfaces

The relationship between the coordination number and bond length may now be applied to crystal surfaces. A crystal surface is characterized by the discontinuation in structural features and hence in properties. The coordination number of a surface atom is smaller than that of an atom within the crystal lattice and hence the application of the above mentioned rules suggests a shortening of the bonds within the surface area. This conclusion should also be applicable to the surface area of a liquid and hence this model may be useful in obtaining a structural interpretation for the phenomenon of surface tension.

It is very interesting that *Lennard-Jones* (*51*) predicted in 1928 that lattice parameters near and perpendicular to the surface should be smaller as compared to those within the crystal lattice, but unfortunately this was not confirmed by experiment (*52*). Indeed truly clean surfaces are hard to obtain, as the highly reactive surface area tends to react with constituents of the atmosphere such as oxygen or water. Under normal conditions many surfaces are covered by oxides or hydroxides, by which the effect of bond contraction at and beneath the surface area is greatly reduced or cancelled. Strong adsorption may even lead to greater bond length than within the bulk of the crystal lattice.

When Boswell (*53*) produced experimental evidence for lattice contraction of NaCl microcrystals in 1951 by means of electron diffraction experiments, no ref-

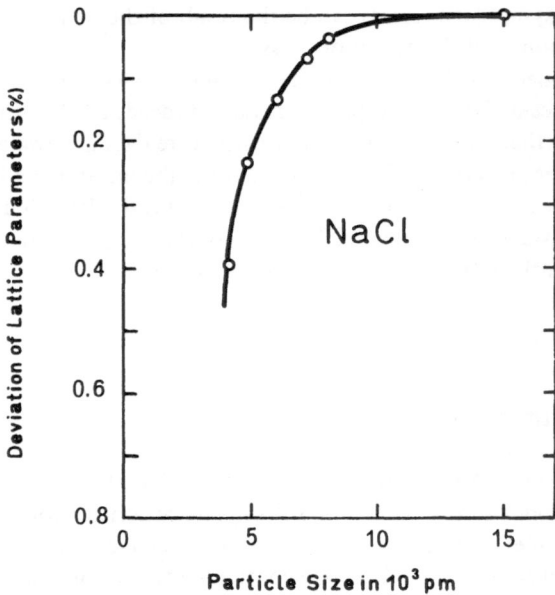

Fig. 1. Decrease in lattice parameters of NaCl crystals with decrease in particle size (*53*).

erence was made to the postulate of Lennard-Jones. The lattice constants were found to decrease with the size of the crystallites below 15000 pm, as shown in Fig. 1. Similar results were obtained on solid silver (*54*).

Further experimental support has been provided by the results of structure determinations of surfaces cleaned by ion bombardment and the annealing method (*55*). They show that the surface atoms are rearranged from bulk positions especially for the diamond structures of the semiconductors Ge and Si. At a clean cleavage plane of silicon the Si-tetrahedra are distorted to such an extent that the interatomic distances between surface Si-atoms are 260 pm as compared to 384 pm within the crystal lattice (*55*).

Meanwhile theoretical considerations were presented by *Koutecky* (*56*). He considers the bonds between the first and the second atom below the surface stronger than within the lattice, but he does not refer to Boswell's experimental results. Likewise *Boehm* (*57*) considers the crystal surface as an area of lattice distortion capable to appreciable lattice deformations.

Further experimental evidence for the lattice contraction at the crystal surface has been obtained from the Mössbauer isomer shift of [197]Au measured in samples of gold microcrystals supported in gelatine (*58*): the experimental isomer shift is correlated with the lattice contractions observed in these microcrystals and the data are shown in Fig. 2.

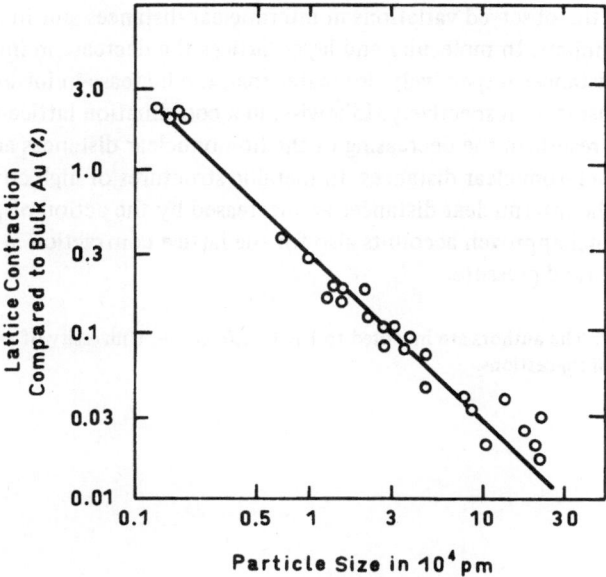

Fig. 2. Lattice spacing in gold microcrystals as a function of the particle size (*58*).

Due to the asymmetric environment of the surface atoms the bonds formed by them are shorter than within the symmetric arrangement in the bulk:

Therefore there will be a surface free energy which is related to the surface tension in the liquid state. For microcrystals the lattice distortions may be considered to be essentially homogeneous throughout the whole microcrystal volume. It has even been suggested that such lattice changes be called "internal pressures" (*58*) (In NaBr crystals of particle size of 5.000 pm the contraction is considered equivalent to an internal pressure of 2 kbar.).

4. Conclusion

The functional approach as recently applied to coordinating and redox reactions, has been used in order to give an interpretation of the variations in internuclear distances observed under pressure in various crystal structures. The action of pressure is considered as paralleling that of an electron donor in that the electron density is increased in internuclear areas of initally low electron density. This formal analogy ac-

63

counts both for the observed variations in internuclear distances and in changes in coordination numbers. In molecular and layer lattices the decrease in intermolecular and interlayer distances respectively, is greater than the increase in intramolecular and intralayer distances respectively. Likewise, in a coordination lattice the application of pressure results in the decreasing of the homonuclear distances and in the increasing of the heteronuclear distances. In metallic structures of high coordination numbers all of the internuclear distances are decreased by the action of pressure.

The functional approach accounts also for the lattice contraction at a crystal surface under normal pressure.

Acknowledgement: The authors are indebted to Dr. J.F. *Jameson*, University of Dundee, for his interest and helpful suggestions.

References

1. *Neuhaus, A.:* Mh. Mineralog. 250 (1963).
2. *Neuhaus, A.:* Chimia *18*, 93 (1964).
3. *Hall, T. H., Barnett, J. D., Merril, L.:* Science *139*, 111 (1963).
4. *Bundy, F. P.:* Science *137*, 1055 (1962).
5. *Wentorf jr., R. H.:* J. chem. Phys. *36*, 1990 (1962).
6. *Alder, B. J., Christian, R. H.:* Phys. Rev. Letters 7, 367 (1961).
7. *Stishov, S. M., Popova, S. V.:* Geochemistry 6, 923 (1961).
8. *Akimoto, S., Syono, Y.:* J. Geophys. Res. *74*, 1653 (1969).
9. Modern, Very High Pressure Techniques (ed. R. H. Wentorf). London: Butterworth 1962.
10. Progress in Very High Pressure Research, Proceed, Int. Conf. New York 1961.
11. Physics and Chemistry of High Pressures, Soc. of Chem. Ing., London, June 1962.
12. *Simmons, G., Bell, P.:* Science *139*, 1197 (1963).
13. *Kleber, W., Wilke, K.-Th.:* Kristall u. Technik *4*, 165 (1969).
14. *Neuhaus, A.:* Geol. Rundschau *57*, 972 (1968).
15. *Klement, W. jr., Jayaraman, A.:* Progr. Solid State Chem. *3*, 289 (1967).
16. *Jamieson, J. C.:* Science *139*, 1291 (1963).
17. *Jamieson, J. C.:* Science *139*, 762 (1963).
18. *Barnett, J. D., Vern, E. B., Hall, H. T.:* J. Appl. Phys. *37*, 875 (1966).
19. *Kleber, W.:* Kristall u. Technik *2*, 13, (1967).
20. *Jørgensen, C. K.:* Molec. Phys. *2*, 309 (1959).
21. *Lüdemann, H. D., Franck, E. U.:* Ber. Bunsenges. physik. Chem. *71*, 455 (1967).
22. *Lüdemann, H. D., Franck, E. U.:* Ber. Bunsenges. physik. Chem. *72*, 514 (1968).
23. *Drickamer, H. G., Frank, C. W.:* Electronic Transitions and the High Pressure Chemistry and Physics of Solids. London: Chapman and Hall 1973.
24. *Gutmann, V.:* Chemische Funktionslehre. Wien–New York: Springer-Verlag 1971.
25. *Gutmann, V.:* Coord. Chem. Rev. *15*, 207 (1975).
26. *Prout, C. K., Wright, J. D.:* Angew. Chem. *80*, 688 (1968).
27. *Ukaji, T., Kuchitsa, K.:* Bull. Chem. Soc. Japan, *39*, 2153 (1966).
28. *Van Bolhuis, F., Koster, P. B., Migchelsen, T.:* Acta Crystallogr. *23*, 90 (1967).
29. *Gutmann, V.:* Electrochim. Acta, *21*, 661 (1976).
30. *Noll, W.:* Angew. Chem. *75*, 123 (1963).
31. *Migchelsen, T., Vos, A.:* Acta Crystallogr. *23*, 796 (1967).
32. *Krebs, H.:* Grundzüge der anorganischen Kristallchemie. Stuttgart: Ferd. Enke Verlag 1968.
33. *Smith, P. L., Martin, J. E.:* Physics Letters *19*, 541 (1965).
34. *Basett, W. A., Takahashi, T.:* Amer. Mineralog. *50*, 1576 (1965).
35. *Busmann, E.:* Z. anorg. Allg. Chem. *313*, 90 (1961).
36. *Zoltai, T., Buerger, M. J.:* Z. Krist. *111*, 129 (1959).
37. *Ringwood, A. E.:* Earth Planet. Sci. Letters *2*, 255 (1967).
38. *Mayer, H.:* Mh. Chem. *105*, 46 (1974).
39. *Tillmanns, E., Gebert, W., Baur, W. H.:* J. Solid State Chem. *7*, 69 (1973).
40. *Bissert, G., Liebau, F.:* Acta Crystallogr. *B 26*, 233 (1970).
41. *Liebau, F., Hesse, K. F.:* Z. Kristallogr. *133*, 213 (1971).
42. *Edge, R. A., Taylor, H. F. W.:* Acta Cryst. *B 27*, 594 (1971).
43. *Flynn, J. J., Boer, F. D.:* J. Amer. Chem. Soc. *91*, 5756 (1969).
44. *Stishov, S. M., Belov, N. V.:* Dokl. Akad. Nauk USSR *143*, 951 (1962).
45. *Preisinger, A.:* Naturwiss. *49*, 345 (1962).
46. *Ringwood, A. E., Reid, A. F., Wadsley, A. D.:* Acta Crystallogr. *23*, 1093 (1967).
47. *Liebau, F.:* Bull. Soc. fr. Mineral. Cristallogr. *94*, 239 (1971).
48. *Mayer, H.:* Dissertation, Techn. Universität Wien 1971.

49. *Wyckoff, R. W. G.:* Crystal Structures Vol. 1, Sec. Edition. New York—London—Sydney: John Wiley & Sons 1965.
50. *Lawson, A. W., Tang, T. Y.:* Phys. Rev. *76*, 301 (1949).
51. *Lennard-Jones, E. J.:* Proc. Roy. Soc. *A 121*, 247 (1928).
52. *Finch, G. I., Wilman, H.:* Erg. exakt. Naturwiss. *16*, 418 (1937).
53. *Boswell, F. W. C.:* Proc. Phys. Soc., *A 64*, 465 (1951).
54. *De Planta, T., Ghez, R., Pinz, F.:* Helv. Phys. Acta *37*, 74 (1964).
55. *Lander, J. J., Morrison, J.:* J. appl. Phys. *34*, 1403 (1963).
56. *Koutecky, J.:* Angew. Chem. *76*, 365 (1964).
57. *Boehm, H.-P.:* Angew. Chem. *78*, 617 (1966).
58. *Schroer, D., Marzke, R. F., Erickson, D. J., Marshall, S. W., Wilenzick, R. M.:* Phys. Rev. *11*, 4414 (1970).

The Shapes of Main-Group Molecules; A Simple Semi-Quantitative Molecular Orbital Approach

Jeremy K. Burdett

Department of Inorganic Chemistry, The University of Newcastle upon Tyne, Newcastle upon Tyne, NE1 7RU, England.

Table of Contents

The angular overlap model, which has been of use in understanding the electronic structure and spectra of transition metal complexes is used to look at the factors which influence the shapes and relative bond strengths in main group systems AB_n ($n = 2 - 7$). Whilst the method is of some interest in itself, the main value of this paper is to show how several molecular orbital effects (ligand-central atom p orbital bond energy, central atom s orbital involvement, and non-bonded interactions) contribute to determine the overall geometry.

67

1. Introduction

There are basically three ways in which the shapes of simple molecules of the main group elements have been explained over the years.

(i) The VSEPR method (*1–5*) was developed by *Sidgwick, Powell, Nyholm* and *Gillespie* and *others* and is summarised in the Nyholm-Gillespie rules. This well known approach considers the geometry to be determined by the minimum energy of the repulsion of electron pairs surrounding the central atom. No really convincing explanation of the far-reaching success of this model in predicting molecular stereo-chemistry, exists on a molecular orbital basis.

(ii) *Walsh* derived (*6*), in qualitative fashion, the change in energy of the molecular orbitals of some simple systems as a symmetrical geometry was deformed. (e.g. planar to pyramidal AB_3 molecules.) Some orbital energies were found to change considerably more than others. On this model, where electron-electron repulsions are explicitly neglected, the change in total orbital energy on changing the geometry, as a function of the number of valence electrons (e.g. BH_3, CH_3, NH_3) gave valuable clues as to the likely geometry a molecule would adopt. Whether the highest oc-cupied molecular orbital (HOMO) of a symmetric geometry changed energy sig-nificantly on deformation was regarded as the major criterion for whether the dis-torted geometry would result. The numbers of electrons needed to send one geo-metry into another (e.g. BH_3 planar, 6 electrons; NH_3 pyramidal, 8 electrons) gave rise to Walsh's rules. *Parr* and *co-workers* have made some useful additional com-ments to the scheme (*8*).

(iii) *Bartell* (*8*) used the second order (or pseudo) *Jahn-Teller* formalism to investi-gate from symmetry considerations alone, whether a symmetrical geometry was stable or unstable with respect to deformation along a given symmetry coordinate. *Pear-son* (*9*) used the method to derive his symmetry rules for molecular shapes. Mathe-matically, the method reduces to a test of whether the HOMO is stable or unstable with respect to a particular deformation mode and thus has strong links with (ii).

Non-bonded interactions have also been shown to determine some of the finer features, and in some cases the gross features of the geometries of molecules (*8, 10*). For example, the planarity of $N(SiH_3)_3$ compared to pyramidal $N(CH_3)_3$ and long the classic example of $(d–p)\,\pi$ bonding, has been convincingly argued (*11*) as being the result of non-bonded repulsions between the SiH_3 groups using the Bartell 'hard-sphere' model.

Simple molecular orbital calculations of the Extended Hückel (EHMO) type have given in general good agreement between observed and calculated geometries (*12, 13*). Especially, the work of *Gimarc* (*14*) has put the Walsh diagrams on a quantitative footing. An analysis of the molecular orbital results in general shows that the cal-culated equilibrium geometries are those where the ligand-central atom overlap is largest for the occupied orbitals. Thus the geometries corresponding to the minimum repulsion energy of electron-pairs around the central atom (VSEPR) is also the one of maximum overlap (and presumably formation of strongest bonds) between central atom and ligands.

For transition metal complexes, (i) cannot universally be applied and (iii) does not seem to work in all cases. *Walsh* type diagrams have been constructed (*15, 16*) for these systems and the observed geometries rationalised as a function of the number of electrons present. A successful qualitative method for looking at transition metal stereo-chemistry (*17*) has been developed, in terms of the angular overlap method (previously used mainly in the analysis of electronic spectra) *based* on ligand-metal d orbital interactions. In this paper we shall use the same method but consider ligand-central atom p and s orbital interactions for the main group systems. We will show that the Walsh diagrams linking structures of different geometries may be derived in a simple fashion using very basic algebra, rather than the services of a computer. Whilst the method cannot rival the VSEPR method's rules of thumb for simplicity, or the complete molecular orbital treatment for a detailed analysis of electronic structure, it sheds more light on the details of the molecular orbital framework and its influence on molecular geometry than the former and provides a simple basis on which to understand the results of the latter. As a ready predictor of relative bond lengths, bond angles and overall molecular structure it could occupy a useful place in structural chemistry.

2. Method

Whilst having been a useful tool in transition metal chemistry (*18–21*), the angular overlap model has not been generally applied to main group systems. The approach assumes the interaction energy between two orbitals ϕ_i, ϕ_j on different centres to be proportional to the square of the overlap integral, S, between the two, viz

$$\epsilon = \beta S_{ij}^2 \qquad (1)$$

β is a constant, inversely dependent on the energy separation

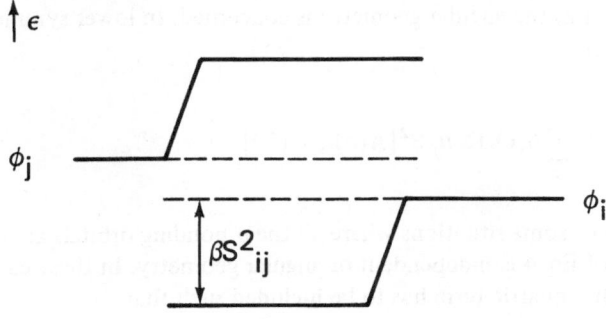

J. K. Burdett

of ϕ_i and ϕ_j and is a measure of the 'strength' of the interaction between the two orbitals. ϵ measures the stabilisation energy afforded to the bonding molecular orbital and although not exactly true we shall also consider ϵ to be equal to the destabilisation energy of the antibonding molecular orbital. In a polyatomic system S may represent a group overlap integral of a set of ligand orbitals with the central atom and the total stabilisation energy of the AB_x system relative to that of its constituent ions can then the written

$$\Sigma = \beta \sum_j n_j S^2 [A(\Gamma_j); L(\Gamma_j)] \tag{2}$$

where n_j is the number of electrons in the bonding orbital of the j-th representation. The overlap integrals S will in general be a product of a constant term (S_σ or S_π depending upon whether the interaction is of the σ or π type) and a function of the angular arrangement of the ligands around the central atom; $f(\Theta, \Phi)$.

$$S = S_\sigma f(\Theta, \Phi) \tag{3}$$

Throughout this paper all bond lengths will be put equal to one another and thus S_σ appears as a constant parameter for all overlap integrals between the central atom and ligand σ orbitals. It is the geometry giving rise to the maximum weighted overlap integral (of Eq. 2) which is the equilibrium geometry of the system. ($f(\Theta, \Phi)$ is in general a readily determinable function involving a little trigonometry as we shall see below. Since π interactions are generally considered to be less important than σ ones in determining molecular geometry and are neglected in the VSEPR scheme we rewrite Eq. 2 as

$$\Sigma(\sigma) = \beta_\sigma \sum_j n_j S^2 [A(\Gamma_j); \sigma(\Gamma_j)] \tag{4}$$

Here we ignore the effect of the angular dependence of any π type stabilisation and assume complete impotence of all electrons located in π orbitals located initially on the ligands as far as the angular geometry is concerned. In lower symmetry situations Eq. 4 becomes

$$\Sigma(\sigma) = \sum_x \beta_\sigma(X) \Sigma n_j S^2 [A(\sigma); \sigma(X)] \tag{5}$$

for all ligands X. For some situations where all the σ bonding orbitals are occupied we find the $\Sigma(\sigma)$ of Eq. 4 is independent of angular geometry. In these cases it can be shown (17) that a quartic term has to be included such that

70

$$\Sigma'(\sigma) = \beta_\sigma \sum_j n_j \, S^2[A(\Gamma_j) \; ; \; \sigma(\Gamma_j)] - \gamma_\sigma \sum_j n_j \, S^4[A(\Gamma_j) \; ; \; \sigma(\Gamma_j)] \qquad (6)$$

where $\beta_\sigma > \gamma_\sigma$ and $\beta_\sigma, \gamma_\sigma > 0$. Thus the equilibrium geometry is the one with the *maximum* weighted (with the electron occupancy of the bonding orbitals) sum of the squares of the overlap integrals, or if this function is angle independent, the equilibrium geometry is determined by the *minimum* weighted sum of the fourth powers of the overlap integrals. We shall use these two criterion below to arrive at the equilibrium geometries of main group compounds, demanded by ligand σ-central atom p orbital interactions. The spherically symmetrical central atom s orbital gives rise to overlap integrals with the ligands which are independent of geometry. However, as we shall see, $s-p$ mixing via a_1 molecular orbitals has a dramatic effect on the molecular geometry.

3. AB₂ Systems

Let us consider two identical ligand σ orbitals pointing toward a central atom $(\sigma_{1,2})$. In the C_{2v} point group we may construct from these, two ligand symmetry adapted combinations.

$$\Phi(a_1) = \frac{1}{\sqrt{2}} (\sigma_1 + \sigma_2) \qquad (7)$$

$$\Phi(b_2) = \frac{1}{\sqrt{2}} (\sigma_1 - \sigma_2)$$

The central atom p orbitals transform as $a_1 + b_1 + b_2$ and thus the molecular orbital diagram for this simple triatomic system takes the form of Fig. 1a or 1b depending upon the relative electronegativities of central atom and ligands. (Or more strictly on their relative ionisation potentials.) Here we have specifically ignored the presence of an s orbital on the central atom. As the angle between the ligand σ orbi-

71

J. K. Burdett

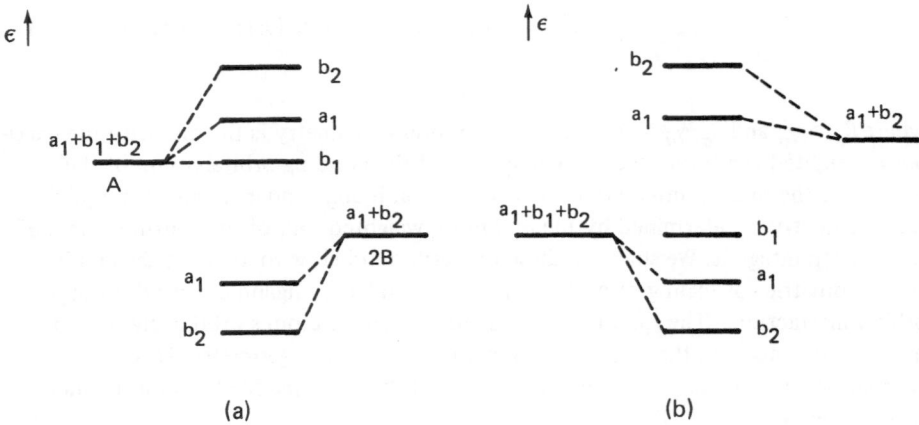

Fig. 1. Molecular orbital diagram for an AB_2 system (a) ligands more electronegative than central atom (b) less electronegative.

tal and the central atom p orbital changes, then the overlap integral between them is given by the simple function $S(\alpha) = S_\sigma \cos \alpha$ (Fig. 2). Using Eq. 1 and the wavefunctions of Eq. 7, the energy dependence upon angle of the a_1 combination is simply given by $\epsilon(a_1) = 2\beta_\sigma S_\sigma^2 \sin^2\alpha$ and of the b_2 combination by $\epsilon(b_2) = 2\beta_\sigma S_\sigma^2 \cos^2\alpha$ (Fig. 3) due to overlap with the relevant central atom p orbital. We now have a semi-quantitative Walsh diagram and we use these calculated energy dependences on angle to calculate the lowest energy (maximum stabilisation energy) geometries using Eqs. 4 or 6 above. It is a general feature (17, 20) of the method that the sum of the quadratic stabilisation energies of Eq. 4 of all the bonding orbitals is equal to $n\beta_\sigma S_\sigma^2$ where n is the number of σ ligands. This serves as a useful check on the arithmetic involved in deriving the ϵ values. The simple Walsh diagram of Fig. 3 shows that the angular energetic behaviour of the antibonding orbitals mirrors that of the corresponding bonding orbitals. As we have noted above, this is only approximately true but since we shall for the most part be interested in the qualitative behaviour of the bonding orbitals only, such inaccuracy will not worry us.

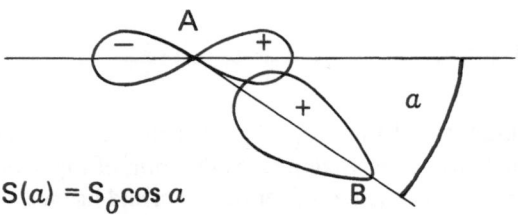

$$S(a) = S_o \cos a$$

Fig. 2. Overlap of ligand σ orbital with central atom p orbital as a function of angle.

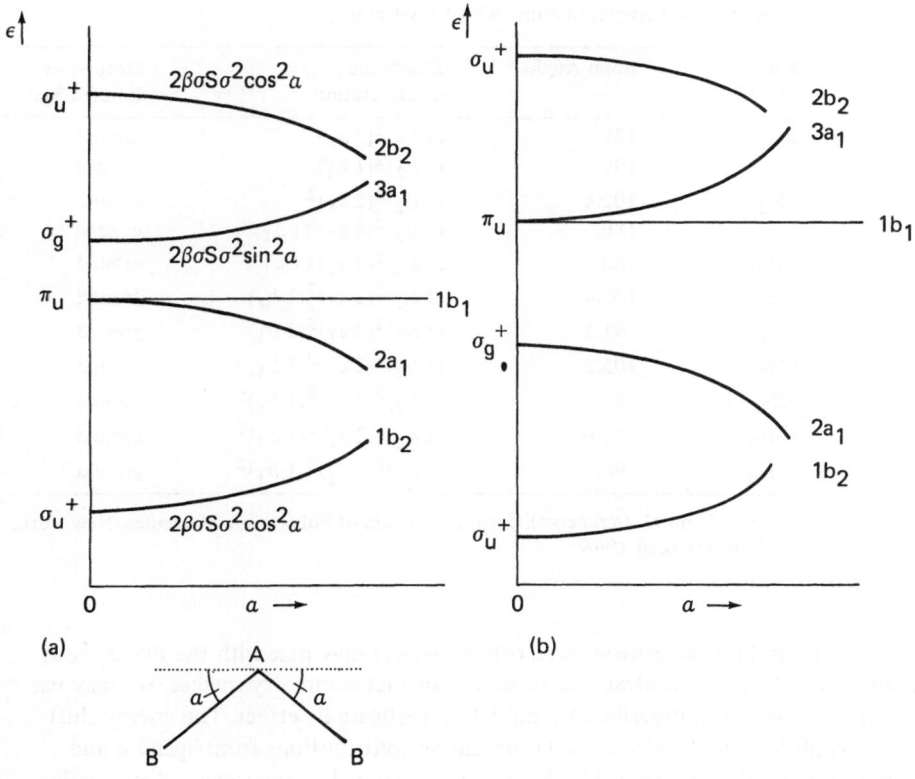

Fig. 3. Molecular orbital energy changes as a function of bending angle for the cases where (a) central atom is more electronegative and (b) where it is less electronegative than the ligands.

For BeH_2 the orbital configuration is $(1b_2)^2$ and thus $\Sigma(\sigma) = 4\beta_\sigma S_\sigma^2 \cos^2\alpha$ with a well defined maximum at $\alpha = 0$. The molecule should therefore be linear. Singlet CH_2, NH_2 and OH_2 all have double occupancy of $2a_1$ and thus $\Sigma(\sigma) = 4\beta_\sigma S_\sigma^2 (\cos^2\sigma + \sin^2\sigma) = 4\beta_\sigma S_\sigma^2$ i.e. a function independent of angle. We therefore need to revert to consideration of the quartic terms of Eq. 6 to evaluate the minimum energy geometry. Their inclusion gives $\Sigma(\sigma) = 4\beta_\sigma S_\sigma^2 - 8\gamma_\sigma S_\sigma^4(\cos^4\alpha + \sin^4\alpha)$. This function has a maximum value at $\alpha = 45°$. We therefore expect all these species to be bent with apex angles of $90°$. Although the first-row molecules noted above are all non-linear systems the observed bond angles are about $15°$ higher than this. For the second row dihydrides, however, bond angles close to $90°$ are found. (Table 1)

We return to this rather interesting point later. For systems with four pairs of p manifold electrons (e.g. XeF_2, ICl_2^-) the $3a_1$ orbital is populated and now $\Sigma(\sigma) = 4\beta_\sigma S_\sigma^2 \cos^2\alpha$ with the expected result that these species are linear ($\alpha = 0$).

Table 1. Bond Angles in Some AH_2 Dihydrides.[a]

Species	Bond Angle	Electronic Configuration	Ground or Excited State
BH_2	131	$(1 b_2)^2(2 a_1)$	ground
	180	$(1 b_2)^2(1 b_1)$	excited
$CH_2(s)$	102.4	$(1 b_2)^2(2 a_1)^2$	ground
	180	$(1 b_2)^2(2 a_1)(1 b_1)$	excited
$CH_2(t)$	180	$(1 b_2)^2(2 a_1)(1 b_1)$	ground
NH_2	103.4	$(1 b_2)^2(2 a_1)^2(1 b_1)$	ground
PH_2	91.5	$(1 b_2)^2(2 a_1)^2(1 b_1)$	ground
CH_2	105.2	$(1 b_2)^2(2 a_1)^2(1 b_1)^2$	ground
SH_2	92.2	$(1 b_2)^2(2 a_1)^2(1 b_1)^2$	ground
SeH_2	91.0	$(1 b_2)^2(2 a_1)^2(1 b_1)^2$	ground
TeH_2	90.2	$(1 b_2)^2(2 a_1)^2(1 b_1)^2$	ground

[a] Data from *G. Herzberg:* Electronic Spectra of Polyatomic Molecules. New York: Van Nostrand 1966.

The s orbital on the central atom (of species a_1) may mix with the two a_1 combinations involving the central atom p orbital of that symmetry species. We may use overlap considerations described by Eq. 1 to investigate its effect. The energy shift of a p manifold a_1 molecular orbital (containing contributions from ligand σ and central atom p orbitals) from this source is proportional to the square of its overlap integral with the central atom s orbital by an interaction described by Eq. 1. This shift will be largest at that geometry where central atom $s-a_1$ molecular orbital overlap is largest, i.e. where the amount of ligand σ character is a maximum. (Central atom s and p orbitals are of course orthogonal.) At the linear geometry the $1 a_1$ orbital (of Fig. 3) correlates with σ_g^+ (of the $D_{\infty h}$ point group) and at the linear structure is a pure ligand σ orbital, there being no central atom p orbital transforming as σ_g^+ in this point group. At this geometry the $3 a_1$ orbital correlates with one component of the π_u pair of pure central atom p orbitals. Thus the s orbital interaction is at a maximum with the $2 a_1$ orbital and identically zero for the $3 a_1$ orbital at the linear geometry. As the molecule bends $2 a_1$ will possess less and $3 a_1$ possess more ligand character and the relative s orbital destabilisation energies of these two orbitals changes (Fig. 4). The $1 a_1$ orbital (predominantly central atom s) will, to a good degree of approximation, remain unchanged in energy since, whatever the geometry it contains the total central atom s-ligand σ stabilisation energy $(n \beta_s S_{\sigma s}^2)$ of all the n ligands. Since the total s orbital destabilisation energy associated with the p manifold group of a_1 orbitals also remains constant as the geometry changes, the loss in destabilisation energy in $2 a_1$ on bending must be matched by an equal gain by $3 a_1$.

The qualitative behaviour of the a_1 orbitals of Fig. 4 has been drawn bearing this in mind. (23). How large the s orbital contribution will be compared to the energy changes involving the orbitals derived purely from the p manifold obviously depends upon (i) the size of the overlap integrals between ligand σ and s (ii) how deeply the s orbital lies in energy below the p orbital bonding orbitals (of approximately the same ionisation potential as the ligand σ orbitals). Thus, as we will see later, s orbital effects are most apparent in those systems where there is a large electronegativity difference between central atom p and ligand σ orbitals, such that the ligand σ orbitals may lie closest to the deep-lying central atom s orbital. The effect of including such mixing on the angular geometry is clearly zero for BeH_2 since no a_1 orbitals derived from the central atom p-ligand σ manifold are occupied; in all the other systems noted above, this mixing favours even more bent geometries in those molecules where the ligand σ orbitals lie deeper than central atom p and less bent geometries where the opposite is true (Fig. 4). As we can readily see, for example, in the former case (Fig. 4b) the occupied $2a_1$ orbital is steeper with s orbital interaction than without. This is an interesting effect and has its analogy in hybridisation terms where the bond angles are considered to increase with increasing s orbital involvement (p^3, 90°; sp^3, 109.5°; sp^2, 120°; sp, 180°).

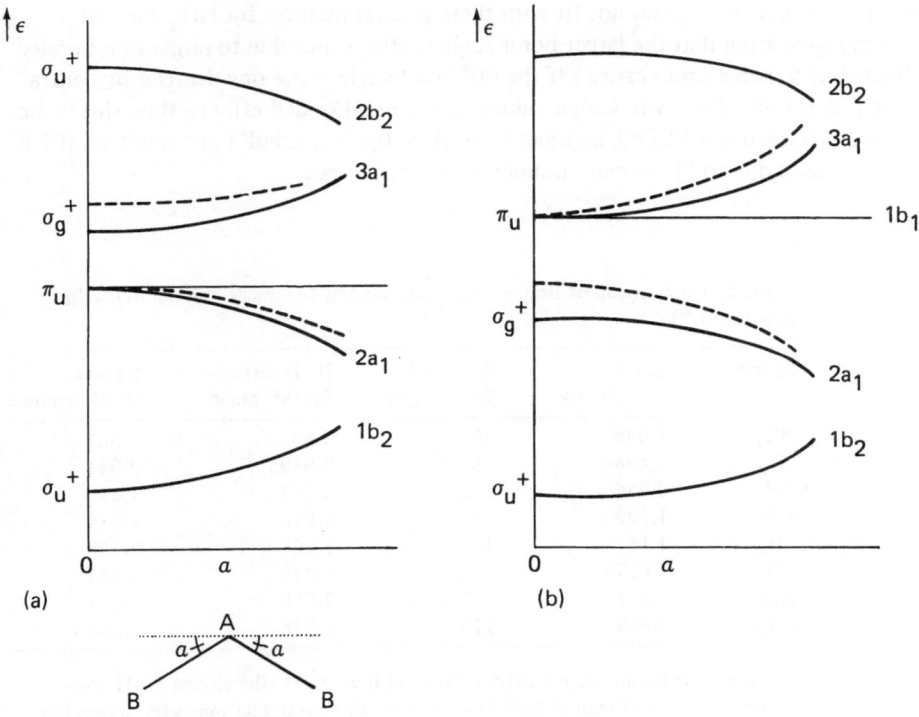

Fig. 4. Effect of including s orbital interaction on the molecular orbital diagrams for AB_2 systems. Dashed lines show where a_1 orbitals have been shifted. (a) Central atom more electronegative and (b) central atom less electronegative than the ligands.

Inclusion of the s orbital then leads to a ready explanation of the bent nature of $BH_2 ((1a_1)^2(2b_2)^2(2a_1)^1)$, pointedly absent from our discussion above. For this species $\Sigma(\sigma) = 2\beta_\sigma S_\sigma^2 (1 + \cos^2\alpha)$ from the p orbital stabilisation and indicates maximum stabilisation of the linear geometry. BH_2, however, is non-linear but has an apex angle approximately halfway between that of a linear molecule and the properly bent examples of Table 1 with the $(2a_1)^2$ configuration. This may represent the balance between the forces of s and p orbital interaction with the ligand σ orbitals. There is more evidence to support this point in the next section concerning CH_3 and we reserve further discussion until then.

The bond angles close to $90°$ predicted for some of these systems are only observed in practice for second row systems (and molecules with even heavier central atoms). One feature of these molecules is that much longer A–H bond lengths are the rule. The larger first row angles could well be due then to non-bonded repulsions between the hydrogen atoms. An analysis using one angle radii to estimate this effect is shown in Table 2. In both cases the H–H distance for a $90°$ geometry is less than the sum of the H one-angle radii, for the first row systems but greater for the second. This indicates that the $90°$ geometry would experience large non-bonded repulsions for the first row systems but not for the rest. (As an aside, we also note that the figures for the BH_2 system do not fit with these general features for NH_2 and OH_2 strongly indicating that the larger bond angle in BH_2 is not due to larger non-bonded effects but to some other cause.) If the $90°$ bond angle is the one decreed by central atom forces but is forced to higher values due to non-bonded effects, then this is the opposite to the usual VSEPR argument (5). Here the 'expected' bond angle of $109.5°$ is compressed due to "lone-pair–bonded-pair" repulsions.

Table 2. Comparison of Bonded and Non-Bonded Distances in AH_2 and AH_3 Systems[a], [b]

Species	Bond Length (A)	Observed Bond angle	H–H distance for 90° angle	Actual H–H distance
NH_2	1.024	103.4	1.448	1.607
PH_2	1.428	91.5	2.0195	2.044
OH_2	0.956	105.2	1.352	1.519
SH_2	1.382	92.2	1.878	1.914
BH_2	1.18	131	1.669	2.148
NH_3	1.0173	107.8	1.439	1.644
PH_3	1.421	93.3	2.010	2.068
CH_3	1.079	120	1.526	1.869

[a] One angle radius for H has been taken as 0.92 A, i.e. the closest H–H non-bonded contact should be 1.84 A. In fact, this parameter may vary somewhat from molecule to molecule. We may expect $r_H(B) > r_H(C) > r_H(N) > r_H(O)$ for example.

[b] Data from G. Herzberg: Electronic Spectra of Polyatomic Molecules. New York: Van Nostrand 1966.

For AB_2 systems (where the ligands contain π orbitals) similar considerations apply to those we have discussed above. NO_2^- (18 electrons) is bent and takes the place analogous to NH_2, NO_2^+ (16 electrons) is linear and takes the analogous place to BeH_2. Very interestingly, NO_2 which should behave similarly to BH_2 has a very similar bond angle (134.1°), again halfway between the bond angles of NF_2 ('typically bent' = 101.0) and the linear configuration. The same argument may be used to rationalise its structure. Π-bonding has been ignored in deriving these bond angles and we have justified its exclusion in determining gross stereochemical features. Not only are the overlap integrals smaller than those involving σ interactions, but the angular variations in overlap integral are also smaller. However, it can readily be appreciated that any π bonding is larger in the linear configuration leading to minima at $\alpha = 0$ from π bonding considerations only. Thus the σ and π weighted geometry of an AB_2 system may be shifted away from the 90° structure. This rationalises the larger angles observed in general for AB_2 systems compared to AH_2 species. (The VSEPR approach called the latter 'anomalously small'.) The inclusion of π bonding also provides an explanation for the observation that the AB_2 geometries often do not seem to be governed by non-bonded interactions. E.g. in SiF_2 the actual non-bonded distance (2.455 Å) and non-bonded distance if the bond angle were 90° (2.25 Å) are both larger than the sum of the non-bonded radii (2.16 Å).

4. AB_3 Systems

In the C_{3v} geometry both the ligand σ orbitals and the central atom p orbitals transform as $e + a_1$. The bonding and antibonding orbitals formed from ligand σ-central atom p orbital interactions only are thus mirror images of each other. By using symmetry adapted ligand σ orbital combinations after the style of Eq. 7, the quadratic energies of the two symmetry types as a function of out of plane angle are readily calculated as $\epsilon(e) = \frac{3}{2} \beta_\sigma S_\sigma^2 \cos^2 \theta$ and $\epsilon(a_1) = 3 \beta_\sigma S_\sigma^2 \sin^2 \theta$. We may thus readily construct a Walsh diagram of Fig. 5 for the out-of-plane deformation coordinate. Addition of s orbital interaction follows an identical argument to that discussed above for AB_2 systems and its effect is similar. For BH_3 and BX_3 the total p orbital stabilisation energy is $6 \beta_\sigma S_\sigma^2 \cos^2 \theta$ with a maximum therefore at $\theta = 0$. BH_3 and BX_3 are thus planar molecules. For $NH_3 (1 e)^4 (2 a_1)^2$, the quadratic term of Eq. 4 is independant of θ and we turn to the quartic term to evaluate the minimum energy angular geometry. This is $- 9 \gamma_\sigma S_\sigma^2 (\cos^4 \theta)$ with a minimum value at $\sin^2 \theta = \frac{1}{3}$, i.e. a geometry where all three central atom ligand bonds make angles of 90° with each other. Again, as in the triatomic case, we find bond angles close to 90° for the second row system PH_3 where the P–H bond is long enough to allow such an angle without large non-bonded repulsions. For NH_3 the bond angle is somewhat larger (Table 2)

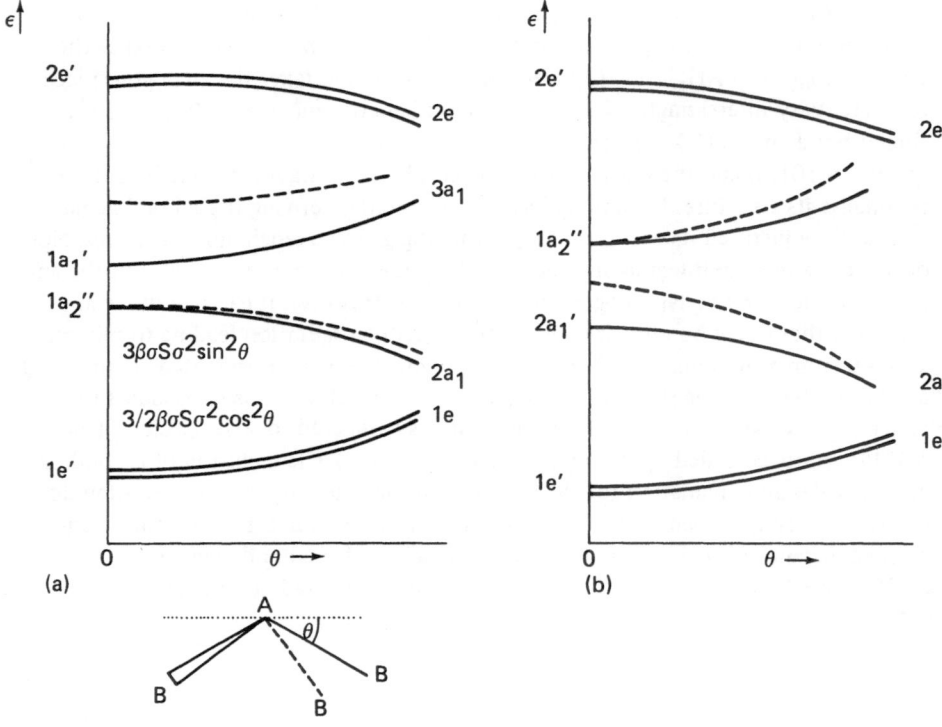

Fig. 5. Molecular orbital diagrams showing the bending of an AB_2 molecule with and without the effect of s orbital mixing (dashed lines). (a) Central atom more electronegative (b) less electronegative than the ligands.

commensurate with the shorter NH bond length. For AB_3 systems π bonding favors the planar geometry and thus fine tuning of the σ determined geometry in these systems is toward larger bond angles. For CX_3 and CH_3 systems, behavior similar to BH_2 and NO_2 might be expected and be dependent upon the degree of involvement of the carbon $2s$ orbital. In fact, CH_3 is planar whereas CX_3 systems are pyramidal (24) to some degree or another. The bond angles of these AH_3 and AB_3 systems are nicely rationalised along the following lines. (The argument is equally applicable to the bending of AH_2 and AB_2 molecules and in fact to the C_{4v} distortion of AB_4 molecules which will be looked at in the next section.) Consider a system where the ligand σ orbitals lie deeper than the central atom p orbitals (Fig. 5a). As the ligand ionisation energy becomes larger, so the interaction between ligand-σ and central atom s becomes bigger. The slope of the a_1 orbitals in both the AB_2 and AB_3 systems then becomes steeper. This implies that bent structures are more favored than linear or planar ones as the ligand electronegativity increases. This is neatly illustrated by e.s.r. results (25, 26) on CH_3, CH_2F, CHF_2 and CF_3 where the degree of bending from planar increases in approximately 5 degree steps from CH_3 (planar) to CF_3 (pyramidal with $\theta = 17.8°$). CCl_3 is also pyramidal (27). Concurrently from Fig. 5a we also see a drop in the amount of s orbital involvement in the p orbital manifold bonding orbitals. These molecular orbital arguments then dovetail well with Mulliken's concept (28) of isovalent hybridisation and Bent's tangent sphere ideas (29) where the more electronegative ligand demands more p character from the central atom. This, in turn, implies a more bent structure. We have previously used (24) arguments along these lines to rationalise the increase in pyramidality of the methyl and substituted methyl radical with increasing total ligand electronegativity. The VSEPR approach rationalises these bond angles by observing that the most electronegative ligands have the smallest bonding pairs and these are most readily squeezed together by the central atom lone pair.

Similar converse arguments hold for the case where the ligand σ ionisation energy is smaller than that of the central atom p orbital. Here increasing electronegativity difference leads to a steeper $2a_1$ orbital favoring the linear or planar configuration for AB_2 and AB_3 systems respectively. It is exceedingly interesting therefore to note in this context that BH_2 is bent (H *more* electronegative than B) whilst CH_3 is planar (C *more* electronegative than H). There are a large number of examples where the bond angles in AB_2 and AB_3 units in more complex molecules are also very prominently dependant upon the ligand electronegativity compared to that of the central atom. For example, a large number of POP bonds in P/O systems lie around $120-130°$ whereas corresponding PSP angles of $100-110$ are found in P/S systems. Similar structural data exists for SOS (114–125) and SSS (100–105) systems. The pseudo *Jahn-Teller* approach (8) anticipates the result of the addition of s orbital interactions to Figs. 4 and 5. As the ligand electronegativity increases, the $\sigma_g^* \rightarrow \pi_u$ energy gap (for an AB_2 system for example) is lowered leading to a larger deformation from the linear geometry and steeper slope of the $2a_1$ orbitals on the pseudo Jahn-Teller scheme. Both of these more subtle effects are borne out in the present treatment.

The simple scheme of Eq. 4 is unable to deal with the problem arising in this system and others due to the presence of more than one a_1 representation contained in the ligand σ orbitals. This will be the case for example if the D_{3h} AB_3 molecule is compressed to a C_{2v} (T shape). Here the central atom p orbitals transform as $a_1 + b_1 + b_2$ where the b_1 orbital (perpendicular to the molecular plane is non-bonding). The ligand σ orbitals transform as $2a_1 + b_2$. The total destabilisation energy of the central atom p orbital is readily calculated but the individual energy shifts of the two ligand a_1 combinations will depend arbitrarily on the choice of ligand a_1 wavefunctions. However, solution of a secular determinant containing ligand σ and central atom p orbitals alone leads to the very general result that all the latent a_1 interaction is contained in one a_1 orbital, all other orbitals of this symmetry species remaining non-bonding. These a_1 orbitals may be mixed however by inclusion of ligand-ligand overlap integrals or by allowing ligand σ-central atom s orbital interaction to occur. We shall ignore the former for the time being and use the ideas of this and the previous section to see what happens to the molecular orbital scheme when an s orbital is in-

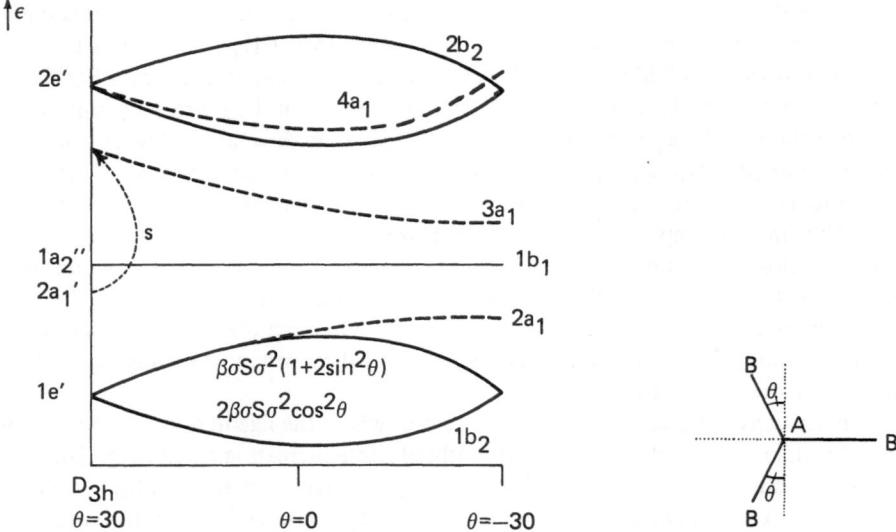

Fig. 6. Correlation diagram for distortion of D_{3h} AB_3 molecule to the arrowhead geometry via a T. Dashed line indicates the orbital behaviour of the a_1 orbitals on bending when s mixing is included.

cluded on the central atom. We remember that the largest interaction will occur with the orbital containing the largest amount of ligand σ character. At the D_{3h} geometry only one orbital involved in central atom p-ligand σ interaction transforms as a_1. Thus all the latent s orbital destabilisation is concentrated here and this orbital, involved in non-bonded interactions with central atom p orbitals is pushed to higher energy (Fig. 6). As the molecule is distorted towards the T shape the degenerate e representations of the ligand orbitals and central atom p orbitals split apart and under the lower symmetry $e \rightarrow a_1 + b_2$. We may readily calculate that for this distortion with no s orbital interaction $\epsilon(b_2) = 2 \cos^2 \theta \, \beta_\sigma \, S_\sigma^2$, $\epsilon(a_1) = (1 + 2 \sin^2 \theta) \, \beta_\sigma \, S_\sigma^2$. Thus the b_2 and a_1 orbitals cross at $\theta = \pm \sin^{-1} \frac{1}{2}$ ($\theta = \pm 30$). This occurs at the D_{3h} geometry and arrowhead $\longrightarrow$ geometry.

Only at the former structure does the lack of s orbital interaction allow them to be degenerate. The destabilisation energy due to s orbital interaction of these a_1 components increases as the distortion from D_{3h} increases, crudely speaking simply because they become less like the e species function at the D_{3h} geometry which has no s orbital interaction. Since the sum total s orbital destabilisation interaction with these a_1 combinations remains constant, if $2a_1$ and $4a_1$ increase in s orbital destabilisation energy then $3a_1$ must decrease in energy to counterbalance this effect. (s–p mixing of this type was a process which Walsh did not consider and thus the T shape molecule is markedly absent from his scheme.)

For systems where the lowest two p manifold orbitals of Fig. 6 are doubly occupied we may readily show that the T shape molecule is not favoured. The quadratic term is independent of angle $(6\beta_\sigma S_\sigma^2)$ but the quartic term is $-4\gamma_\sigma S_\sigma^4[4\cos^4\theta + 4\sin^4\theta + 4\sin^2\theta + 1]$. This has a minimum value at $\theta = 30$, so that D_{3h} and arrowhead geometries are equally favoured on ligand σ–p orbital interactions alone. Inclusion of s orbital interaction in Fig. 6 shows destabilisation of $2a_1$ on bending and thus the D_{3h} geometry is even more favoured and the arrowhead structure positively excluded. BX_3 systems do not distort from D_{3h} therefore. Similar considerations apply to the four valence pair case of NX_3. For ClF_3 (five valence pairs) the $3a_1$ orbital of Fig. 6 is occupied[1]) and the prominent feature associated with it is its rapid stabilisation on distortion to the T shape as we have discussed above. We may readily view the combined effect of the two filled a_1 (p manifold) orbitals $2a_1$ and $3a_1$ on distortion by remembering that the total a_1 stabilisation of all a_1 orbitals remains approximately constant throughout the distortion. Thus the maximum *stabilisation* energy of the configuration $(2a_1)^2(3a_1)^2$ occurs where the *destabilisation* of $4a_1$ is a maximum, i.e. at the arrowhead geometry. The exact angle will be a balance of the energy of the orbitals concerned and non-bonded repulsions which would be expected to disfavor the exact arrowhead geometry with $\theta = -30$. The tilting of the two 'axial' ligands towards the unique ligand viewed in simple molecular orbital terms above, is rationalised under the VSEPR approach as being due to the repulsive effect of two lone pairs on the central Cl atom. Here we see that on a molecular orbital basis the geometry is due to the balance between maximum ligand-central atom p orbital stabilisation and minimum s orbital destabilisation subject to the requirement

---

[1]) This is assuming that $3a_1$ lies above $1b_1$. That this is true is indicated by most molecular orbital calculations.

that the ligands cannot get too close to one another. As noted above, ligand σ-central atom s orbital interactions should be larger the closer the ionisation energies of these two orbitals. Thus, the larger the difference in ligand electronegativity, the larger this interaction. Similarly ligand σ-central atom p orbital interaction will decrease with increasing ligand electronegativity. Thus on moving from ClF_3 to BrF_3 we expect to see increased s orbital involvement and smaller p orbital interaction. This would result in a steeper slope of the $3a_1$ for BrF_3 than for ClF_3 and a larger distortion away from the orthogonal T shape. In agreement with this the observed bond angles are 86 and 87.5 for BrF_3 and ClF_3 respectively although the difference between the two is hardly large.

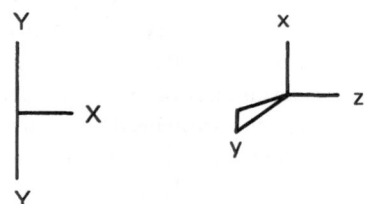

We may also use the theory to investigate the relative bond lengths for the two types of ligand in the T shape geometry by calculating the stabilisation energy associated with each of them. This method is perfectly general for all the systems we shall look at in this paper and one we shall make extensive use of. Consider the orthogonal T structure for simplicity. The unique ligand (X) is bound (in the absence of s orbital terms) by interaction with the p_z orbital and the stabilisation energy associated with it is $2\beta_\sigma S_\sigma^2 - 2\gamma_\sigma S_\sigma^4$ (from Eq. 5). The pair of ligands (Y) forming the arms of the T interact only with p_x and the total stabilisation associated with them is $4\beta_\sigma S_\sigma^2 - 8\gamma_\sigma S_\sigma^4$ (i.e. $2\beta_\sigma S_\sigma^2 - 4\gamma_\sigma S_\sigma^4$ per ligand). This pair of ligands are then individually bound less strongly than the unique one. In complete agreement with this the bond lengths in the two molecules are: BrF_3, 1.81 Å (axial), 1.72 Å (equat) and ClF_3, 1.698 (axial), 1.598 (equat). This is a very much cleaner approach than that of the VSEPR method which is concerned with the notion that lone pair-axial bonded pair repulsions are larger than lone-pair — equatorial bonded pair repulsions. It is also a rationalisation which does not involve the use of d orbitals on the central atom. The approach bears a close resemblance to the results of *Musher's* Hypervalent Molecule Classification (*30*) in which the properties of molecules containing two identical monovalent ligands bound in different ways to a central atom are examined. Thus, for example in the T shaped ClF_3 molecule, two of the fluorine atoms are considered to be bound to the central atom via a three centre — two electron bond involving a single p orbital on the chlorine. The unique ligand is then bound to a single chlorine p_z orbital. The algebraic treatment of the systems described in this paper really reduces to a quantitative treatment (via the relationship of Eq. 1) of the three centre and the two centre bonding situations in these systems.

5. AB₄ Systems

We will consider three geometries in this section, C_{4v} (and the special case of square planar), C_{3v} (and the special case of tetrahedral) and C_{2v} (the disphenoid or 'SF₄' geometry). For the C_{4v} geometry the central atom p orbitals transform as $e + a_1$ and the ligand σ orbitals as $e + a_1 + b_1$. $\epsilon(e)$ is readily calculated as $\epsilon(e) = 2\beta_\sigma S_\sigma^2 \cos^2\theta$ (Fig. 7) and $\epsilon(a_1) = 4\sin^2\theta\,\beta_\sigma S_\sigma^2$ where θ is the droop angle from planar. The molecular orbital diagram constructed using these functions is shown in Fig. 7. For the C_{3v} geometry the p orbitals transform as $a_1 + e$ and the ligand σ orbitals as $2a_1 + e$. We may easily calculate $\epsilon(a_1) = (1 + 3\sin^2\theta)\beta_\sigma S_\sigma^2$ and $\epsilon(e) = \frac{3}{2}\beta_\sigma S_\sigma^2 \cos^2\theta$ where θ is the droop angle from planar of the three symmetry related ligands. The energies are equal at the tetrahedral geometry ($\theta = \sin^{-1}\frac{1}{3}$) where $a_1 + e \rightarrow t_2$ (Fig. 8). For the C_{2v} geometry the ligand σ orbitals transform as $2a_1 + b_1 + b_2$ and the central atom p orbitals as $a_1 + b_1 + b_2$. The interaction energies in the absence of s orbital involvement become $\epsilon(b_1) = 2\beta_\sigma S_\sigma^2 \cos^2\alpha$, $\epsilon(b_2) = 2\beta_\sigma S_\sigma^2 \cos^2\beta$ and $\epsilon(a_1) = 2\beta_\sigma S_\sigma^2 (\sin^2\alpha + \sin^2\beta)$. A schematic molecular orbital diagram containing these energy dependances is shown in Fig. 9.

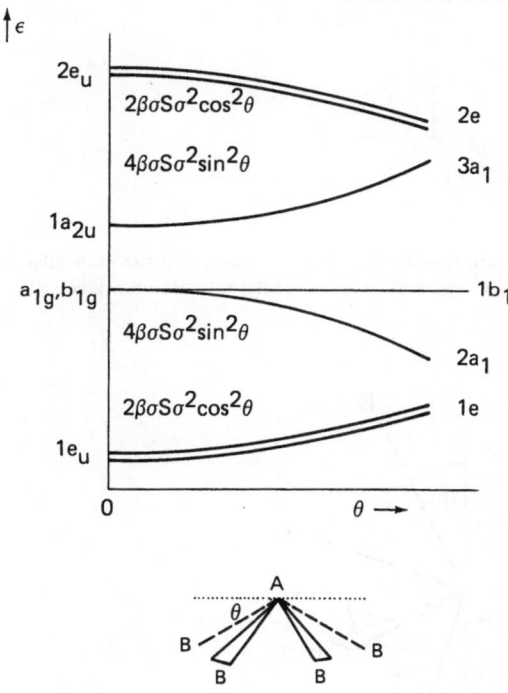

Fig. 7. Molecular orbital diagram showing the change in energy of the molecular orbitals of an AB₄ unit on bending within the C_{4v} coordinate. (s orbital interaction neglected.)

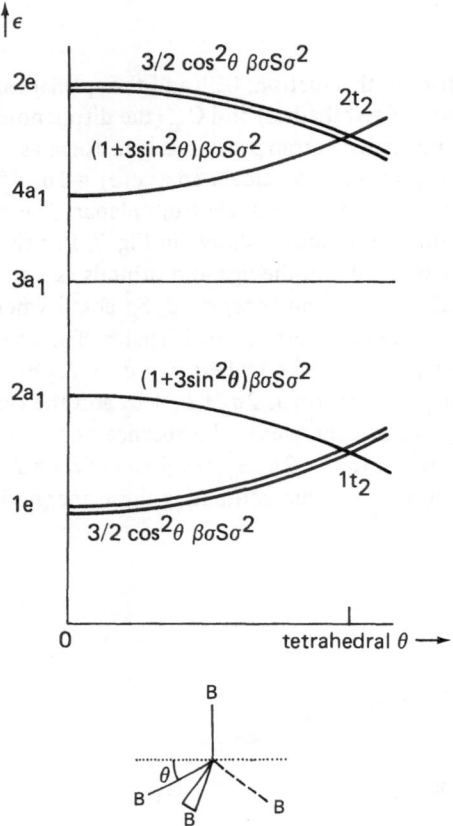

Fig. 8. Molecular orbital diagram showing the change in energy of the molecular orbitals of an AB$_4$ unit on bending within the C_{3v} coordinate (s orbital interaction neglected).

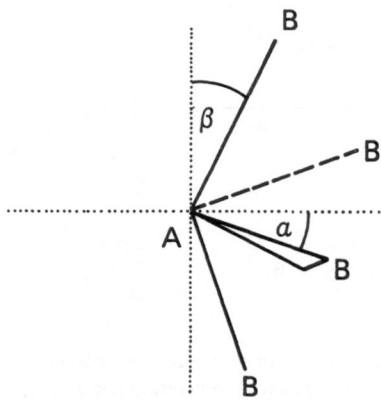

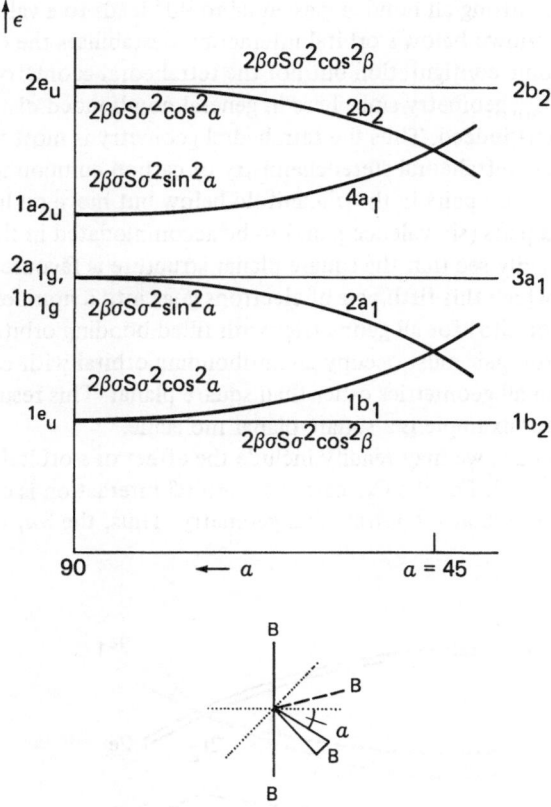

Fig. 9. Molecular orbital diagram showing the change in energy of the molecular orbitals of a square AB_4 unit on bending to the C_{2v} geometry. (s orbital interaction neglected.) $\cos \beta = 0$ for simplicity.

Consider first a system with two electron pairs occupying these orbitals (three valence pairs if a central s orbital is included). $\Sigma(\sigma)$ then becomes $4\beta_\sigma S_\sigma^2$ (square planar), $\frac{8}{3}\beta_\sigma S_\sigma^2$ (tetrahedral) and $4\beta_\sigma S_\sigma^2$ (tetrahedral) and $4\beta_\sigma S_\sigma^2 (\cos^2\alpha + \cos^2\beta)$ for the SF_4 geometry. (For C_{4v} and C_{3v} geometries in general $\Sigma(\sigma)$ is $4\beta_\sigma S_\sigma^2 \cos^2\theta$ and $3\beta_\sigma S_\sigma^2 \cos^2\theta$.) Such a molecule should therefore be square planar. For systems with three p manifold pairs (four valence pairs) e.g. CH_4, CX_4 the quadratic stabilisation energies for all three systems are equal and we turn to a discussion of the quartic terms to pinpoint the equilibrium geometry. For the square planar structure this term is $-16\gamma_\sigma S_\sigma^4$ with a minimum value within the C_{4v} coordinate of $-\frac{32}{3}\gamma_\sigma S_\sigma^4$ at an angle of $\theta = \sin^{-1}\frac{1}{\sqrt{3}} = 35°$. For the tetrahedral geometry the quartic term is $-\frac{32}{3}\gamma_\sigma S_\sigma^4$ which is the smallest value of the quartic term within the C_{3v} coordinate.

85

For the SF_4 structure, putting all bond angles equal to $90°$ leads to a value of $-12\gamma_\sigma S_\sigma^4$. As will be shown below s orbital interaction destabilises the C_{4v} geometry with this electronic configuration but not the tetrahedral geometry. One could also argue against the C_{4v} geometry since here in general non-bonded distances at $\theta = 35$ will be very short indeed. Thus the tetrahedral geometry is most favoured and is found of course in the tetrahedral stereochemistry of carbon compounds. We examine the case of four electron pairs in the p manifold below but move on here to the case of five p manifold pairs (six valence pairs) to be accommodated in the system (e.g. XeF_4). We can readily see that the square planar structure is favoured since it is the only geometry in which this fifth pair of electrons goes into a non-bonding orbital (a_{2u}). Since the value of $\Sigma(\sigma)$ for all geometries with filled bonding orbitals is $8\beta_\sigma S_\sigma^2$ this fifth electron pair must occupy an antibonding orbital with commensurate decrease in $\Sigma(\sigma)$ in all geometries other than square planar. This result is borne out in practice. XeF_4 for example is a square planar molecule.

As in previous systems, we may readily include the effect of s orbital interaction in the schemes of Figs. 7–9. For the C_{3v} case the s orbital interaction is concentrated in the single ligand a_1 orbital at the tetrahedral geometry. Thus, the $3a_1$ orbital re-

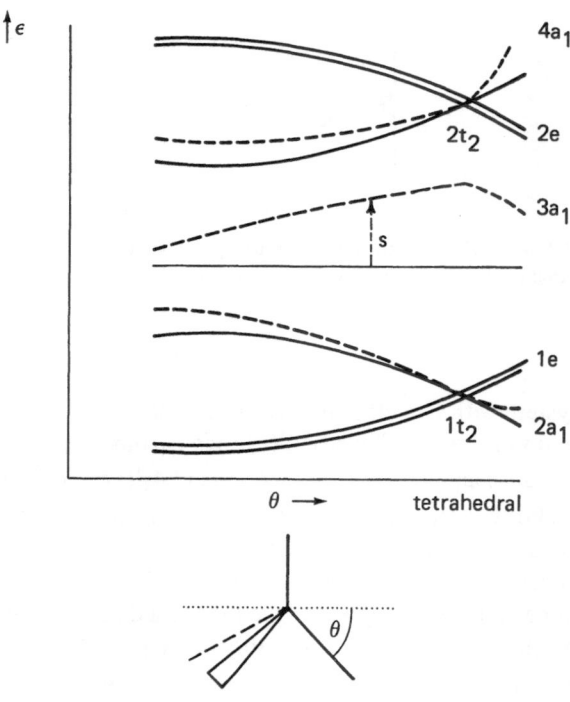

Fig. 10. Part of the molecular orbital diagram showing orbital energies in the C_{3v} distortion coordinate, with s orbital interaction included (dashed lines).

ceives its maximum destabilisation here (Fig. 10). As the geometry is deformed to C_{3v} this s orbital destabilisation energy must now be shared with the $2a_1$ and $4a_1$ orbitals derived from $1t_2$ and $2t_2$ at the tetrahedral configuration. Again, since the total s orbital destabilisation energy of these a_1 orbitals remains constant during this distortion the increase in energy of $2a_1$ and $4a_1$ from this source must be matched by an equal and opposite decrease in $3a_1$. Inclusion of s orbital mixing further underlines the stability of the tetrahedron with four valence pairs. Distortion either side of tetrahedral leads to s orbital destabilisation of the occupied $2a_1$ orbital. We can however see that on the same argument the presence of two more electrons (in $3a_1$) make the tetrahedral geometry unlikely for this electronic configuration. Just as occupation of the $3a_1$ orbital in ClF_3 led to rapid destabilisation of the symmetric D_{3h} geometry so occupation of $3a_1$ in the tetrahedral geometry will lead to destabilisation of this geometry. The amount of s orbital involvement as we have mentioned above will be a function of the nature of the ligand and central atom electronegativity. However, a general feature of molecular orbital calculations on these molecules is that the a_1 orbitals (with the exception of $1a_1$) are the ones which change most in energy on distortion (*31*). Of the a_1 orbitals the most sensitive one of these to distortion is the orbital which receives the maximum s orbital destabilisation at the symmetric geometry (in our cases the $3a_1$ orbital).

In principle we could use the approach developed in this paper to investigate the changes in bond angle expected in the tetrahedral geometry for four valence pairs when the ligands are not all identical, e.g. CF_2Cl_2. However, the observed bond angles are only very slightly different from the tetrahedral ones and we suspect that a subtle blend of factors will influence the angle adopted. It is however profitable at this stage to look at the larger angle changes often introduced when formal double or triple bonds exist between the central atom and one or more of the ligands. The unique ligand is generally oxygen (e.g. the POX_3 series) but may be sulfur (e.g. PSX_3) or nitrogen (e.g. NSF_3).

In these cases the XAX angles are less than tetrahedral. A similar effect pertains in NO_2Cl and allied planar molecules where the angle between the oxygen atoms ($130°$) is larger than that of the trigonal planar geometry. The VSEPR approach attributes the effect to large repulsions by the electrons, in the multiple bond holding the oxygen to the central atom with the other bonding pairs. We may also understand this result however in molecular orbital terms. The noticeable feature concerning

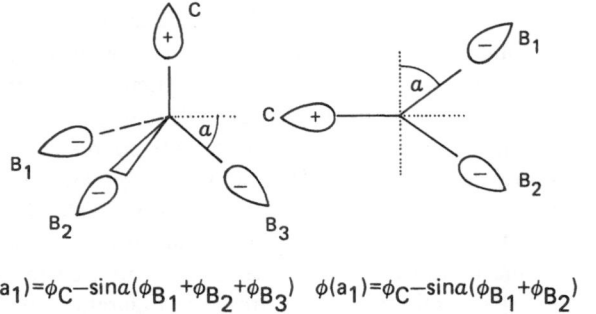

$$\phi(a_1)=\phi_C-\sin\alpha(\phi_{B_1}+\phi_{B_2}+\phi_{B_3}) \qquad \phi(a_1)=\phi_C-\sin\alpha(\phi_{B_1}+\phi_{B_2})$$

Fig. 11. a_1 wavefunctions for tetrahedral and trigonal molecules as a function of distortion.

these multiply-bonded systems is the much shorter A—O bond length than that seen in formally single bonded systems. As a result, we expect larger interaction between central atom p orbitals and ligand σ orbitals than otherwise expected and more importantly in this lower than tetrahedral (or trigonal planar) geometry larger interaction with the central atom s orbital than would be expected on the grounds of ligand electronegativity. In the C_{3v} geometry it is an a_1 orbital derived from the t_2 representation of T_d which receives some of the s orbital destabilisation energy and in C_{2v} it is the a_1 orbital derived from e' of D_{3h}. The ligand σ combinations involved in orbitals of these symmetries are shown in Fig. 11 and the coefficients change with angle as shown. We will require the equilibrium geometry to be that determined by zero s orbital destabilisation of these a_1 orbitals, and therefore need to calculate the overlap integrals between these orbitals and the central atom s orbital. These, as readily seen from Fig. 11 are given by;

$$\text{For } AB_3C \quad S = S_c - 3 \sin \alpha \, S_B$$
$$\text{For } AB_2C \quad S = S_c - 2 \sin \alpha \, S_B \tag{8}$$

For three (or four) equivalent ligands the s orbital overlap (and hence s orbital destabilisation energy) is zero for $\alpha = \sin^{-1}\frac{1}{2}(30°)$ and $\sin^{-1}\frac{1}{3}$ (tetrahedral) respectively. If ligand C has a larger s orbital overlap integral with the central atom s orbital than B then $\sin \alpha$ must increase to keep the total s orbital destabilisation zero. (Thus in POX_3 $\alpha > \alpha_{\text{tet}}$.) Similarly in the case where B has a larger s orbital interaction than C then $\sin \alpha$ must decrease. (Thus in NO_2Cl $\alpha = 25°$.)

For the square planar geometry (Fig. 12) the equivalence in energy of α_{1g} and b_{1g} ligand orbitals is removed as the a_{1g} alone is destabilised by interaction with the central atom s orbital. In many ways the diagram is similar to that for AB_2 systems ($D_{\infty h} \to C_{2v}$) and AB_3 systems ($D_{3h} \to C_{3v}$). If the s orbital interaction is large enough to move the $2a_1$ orbital above the non-bonding $1b_1$ orbital for all values of θ then the C_{4v} geometry is less stable than tetrahedral for the four valence pair case

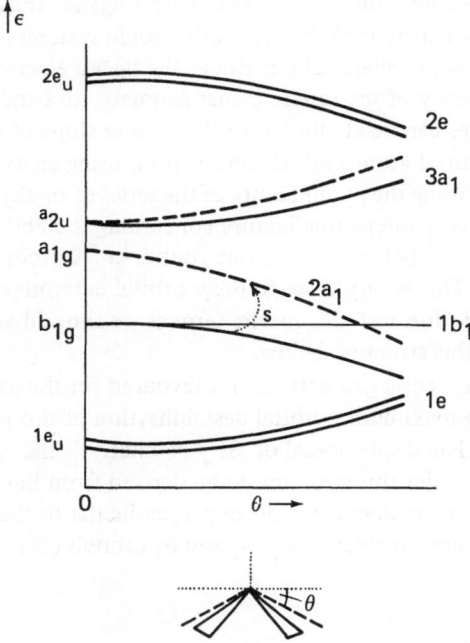

Fig. 12. Molecular orbital energy levels for a C_{4v} distortion of an AB_4 unit showing s orbital interaction.

simply because with this configuration $(1\,e)^4 (1\,b_1)^2$, $\Sigma(\sigma) = 8\beta_\sigma\, S_\sigma^2\, \cos^2\theta$. At best this is smaller than $\Sigma(\sigma)$ for the tetrahedral geometry. If s orbital interaction is smaller such that $2a_1$ may lie below $1\,b_1$ such a geometry is still less favoured than tetrahedral since in the latter *no s* orbital destabilisation occurs at all.

We may use Fig. 12 to investigate what has been called (*32*) the "inert pair effect". The SF_4 geometry and square planar geometries have similar values of the quartic term for four valence pairs and upwards and it is therefore not surprising that some systems exist with local C_{4v} symmetry. It is generally found that in solid PbO and SnO the four oxygen atoms coordinating the metal lie at the corners of a tetra-

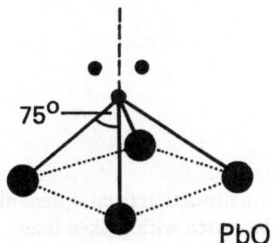

PbO

89

gonal pyramid; but in systems containing less electronegative ligands the coordination is regular (e.g. PbS has the NaCl structure). Even in rather ionic systems (e.g. Tl^I halides) the lone pair become stereochemically active as the ligand electronegativity increases. The increasing tendency of the square planar geometry to bend as the ligand electronegativity increases can be attributed to the steeper slope of the $2a_1$ orbital due to larger ligand σ – central atom s orbital interactions using an exactly analogous argument to that concerning the pyramidality of the series of methyl and substituted methyl radicals. One very interesting feature concerning the PbO structure is that OPbO 'bond angles' of 75° (4) and 115 (2) are found, i.e. A droop angle from planar of the C_{4v} unit of 31°. This is very close to the p orbital determined droop angle of 35° using the algebraic form of the quartic term as we showed above. We shall have more to say about this structure below.

We have seen that the tetrahedral geometry is not favoured for the case of five valence pairs (e.g. SF$_4$) due to maximum s orbital destabilisation of this geometry for this electronic configuration. The disphenoidal or 'SF$_4$' geometry is the one most commonly found. We may consider this structure to be derived from the square plane by bending two central atom ligand bonds in a plane perpendicular to the other two (Fig. 13). The energy dependence on angle of the b_1 and b_2 orbitals (derived from

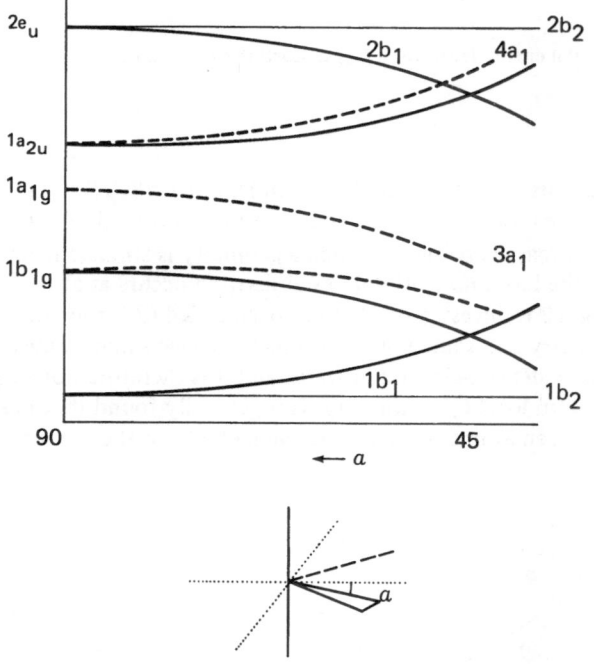

Fig. 13. Molecular orbital energy levels for an AB$_4$ unit on distortion from square planar to the disphenoidal geometry (the effect of s orbital mixing is indicated with dashed lines).

e_u of D_{4h}) and the a_1 orbital (derived from a_{1g}) at D_{4h} may be calculated in the usual way. $\epsilon(b_2) = 2\beta_\sigma S_\sigma^2$, $\epsilon(b_1) = 2\beta_\sigma S_\sigma^2 \cos^2\alpha$ and $\epsilon(a_1) = 2\beta_\sigma S_\sigma^2 \sin^2\alpha$. The correlation diagram takes the form of Fig. 13. At the orthogonal geometry ($\alpha = 45°$) the quartic term of a system with occupation of the lowest three p manifold orbitals is $-12\gamma_\sigma S_\sigma^4$ which is larger than that for the tetrahedral geometry ($-\frac{32}{3}\gamma_\sigma S_\sigma^4$). Also inclusion of s orbital interaction raises the energy of the $2a_1$ orbital and thus de-stabilises the system even more compared to tetrahedral. In the latter geometry there is no such destabilisation of the completely filled $1t_2$ set of orbitals. An alternative geometry to tetrahedral is also the C_{4v} one where the quartic term is smaller ($-\frac{32}{3}\gamma_\sigma S_\sigma^4$ at $\theta = 35°$, $-11\gamma_\sigma S_\sigma^4$ at $\theta = 30°$) than that for the orthogonal dis-phenoidal geometry. A comparison of Figs. 12, 13 suggest that the s orbital destabili-sation energy is probably larger for C_{2v} than C_{4v} geometries for these electronic con-figurations. On both counts therefore the C_{4v} geometry ought to be favoured on electronic grounds. However, for a droop angle of $30°$ the closest F–F non-bonded distance in the C_{4v} structure for an SF_4 molecule (bond length ~ 1.5 Å) is about 1.82 Å. The sum of the one angle radii for the two F atoms is 2.16 Å, thus making this C_{4v} geometry extremely unlikely. This being the case, what then are the criteria for stabilisation of the C_{4v} form relative to C_{2v}? Obviously we need a long A–B bond (to minimise B–B non-bonded contact distances) which will be more likely when A lies at the bottom of the periodic table. B needs to be small to minimise the mini-mum tolerable B–B non-bonded distance. O and F are likely candidates and we may probably exclude larger ligands such as Cl and Br. What we have just described in fact are just the conditions for observation of the (not so) "inert-pair" compounds which adopt the C_{4v} geometry with large droop angles.

Fig. 13 shows that the slope of the $3a_1$ orbital energetically favours increasing α. Similarly, tilting of the two axial ligands toward the equatorial ones should (by sym-metry) also be a favoured process, by reducing s orbital destabilisation of $3a_1$. The explanation of the non-orthogonal structure of SF_4 is thus due to the same molec-ular orbital effect which causes similar tilting in ClF_3. Excessive bending of the axial ligands is prevented, as in ClF_3 by non-bonded repulsions between axial and equatorial ligands.

The relative 'axial' and 'equatorial' bond lengths are readily examined using the same method employed for ClF_3. With an axial-axial angle of $180°$ the axial ligands are attached only to p_x and the total axial stabilisation energy is $4\beta_\sigma S_\sigma^2 - 8\gamma_\sigma S_\sigma^4 (2\beta_\sigma S_\sigma^2 - 4\gamma_\sigma S_\sigma^4$ per ligand) whereas the equatorial ligands are bound to p_x and p_y with a total stabilisation energy of $4\beta_\sigma S_\sigma^2 (\cos^2\alpha + \sin^2\alpha) - 8\gamma_\sigma S_\sigma^4 (\cos^4\alpha + \sin^4\alpha)$. In the latter case the stabilisation energy per ligand is $2\beta_\sigma S_\sigma^4 - 4\gamma_\sigma S_\sigma^4 (\cos^4\alpha + \sin^4\alpha)$. The trigonometrical function in parentheses is always less than 1 (for $\alpha \neq 90$) and so the equatorial ligands are bound more tightly than the axial ones. This is re-flected in the observed geometries where the axial bond lengths are always longer than the equatorial ones (e.g. 1.545 Å (equatorial and 1.646 Å (axial) in SF_4). The strongest σ ligands will always prefer the site with the largest latent bond strength since here their contribution to the overall stabilisation energy is maximised. (We have

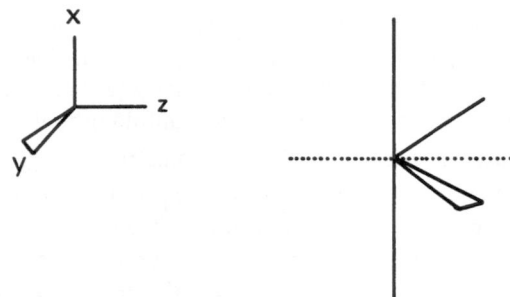

used arguments along these lines in the transition metal complex situation (*33*). Since the more electronegative the ligand, the further apart the ionisation potentials of central atom and ligand orbitals, and the smaller is $\beta_\sigma \, S_\sigma^2$. This implies that the most electronegative ligands prefer the axial positions, as is universally found to be the case in practice. A similar argument follows from the discussion in the previous section concerning the site preferences in the ClF_3 geometry.

6. AB_5 Systems

The two geometries of interest here are the trigonal bipyramid and the square based pyramid, interconvertible via the well-known Berry process. The orbital diagram for the *tbp* geometry is readily constructed (Fig. 14) using our molecular orbital method. Several results from previous sections are immediately transferable to this structure and may be readily proven. The total stabilisation energy of the axial ligands is $4\beta_\sigma \, S_\sigma^2 - 8\gamma_\sigma \, S_\sigma^4 (2\beta_\sigma \, S_\sigma^2 - 4\gamma_\sigma \, S_\sigma^4$ per ligand) and that for the equatorial ligands is $6\beta_\sigma \, S_\sigma^2 - 9\gamma_\sigma \, S_\sigma^4 (2\beta_\sigma \, S_\sigma^2 - 3\gamma_\sigma \, S_\sigma^4$ per ligand). Thus the equatorial ligands are bound more strongly than the axial ones as is found for all known cases. This also implies that the ligands with the largest $\beta_\sigma \, S_\sigma^2$ values will prefer the equatorial sites. This means that the most electronegative ligands should occupy the axial positions (as is the case in PF_3Cl_2, PF_3Br_2, $PF_3(CH_3)_2$ etc.).

For the square pyramidal geometry the molecular orbital diagram is given in Fig. 15. The total stabilisation energy associated with the axial ligand is (at the orthogonal geometry where angle $B_{ax} \, AB_{eq} = 90°$) $2\beta_\sigma \, S_\sigma^2 - 2\gamma_\sigma \, S_\sigma^4$ and with the equatorial ligands is $8\beta_\sigma \, S_\sigma^2 - 16\gamma_\sigma \, S_\sigma^4 (2\beta_\sigma \, S_\sigma^2 - 4\gamma_\sigma \, S_\sigma^4$ per ligand). Thus in structures where the lowest three p orbital manifold orbitals are occupied (four or more valence pairs) as in all known cases, the axial bond length should be shorter than equatorial, as found experimentally, with the possible exception of SbF_5^{-2}. The most electropositive ligand will thus occupy the axial site. The VSEPR approach says that the bonds ad-

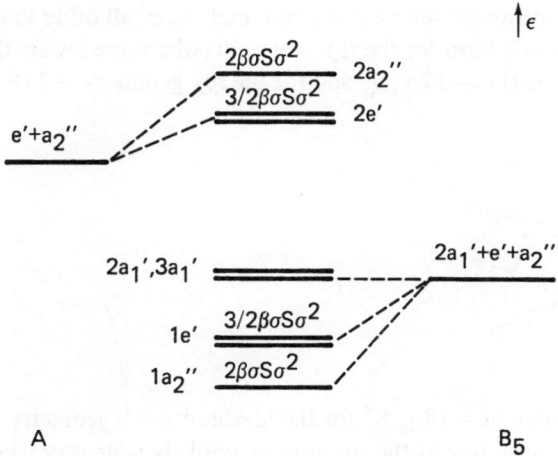

Fig. 14. Molecular orbital energy levels for an AB_5 molecule with the trigonal bipyramid geometry. (No s orbital interaction.)

jacent to the lone pair (equatorial) suffer the greatest repulsion and are hence longer. The present simple explanation is conceptually more reasonable in molecular orbital terms and is also a readily calculable effect.

For the five valence pair case (e.g. PF_5) the two geometries, *tbp* and *spy* are very close in energy as evidenced by the rapid equatorial-axial exchange via an intramolecular process, and the presence in the crystalline state of examples of both structures

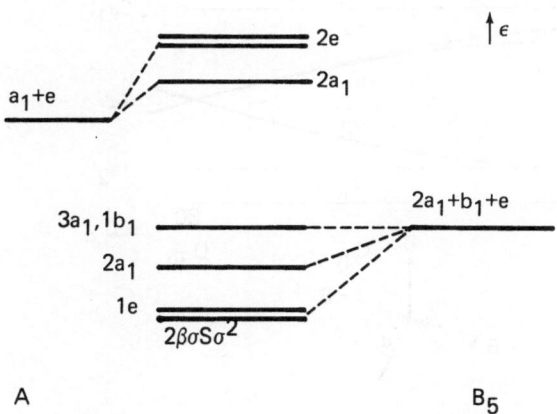

Fig. 15. Molecular orbital energy diagram for an AB_5 molecule with the square pyramidal geometry. (No s orbital interaction.)

93

(SbΦ$_5$ and InCl$_5^{-2}$ are distorted square pyramidal molecules, all other known structures are *tbp*). The quartic term for the *tbp* structure (where the lowest three *p* manifold orbitals are occupied) is $-17\gamma_\sigma S_\sigma^4$ and for the *spy* geometry $-2\,(8\,\cos^4\theta +$
$(1+4\sin^2\theta)^2)\,\gamma_\sigma S_\sigma^4$;

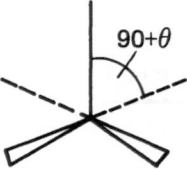

The latter has a value of $-18\gamma_\sigma S_\sigma^4$ for the idealised $\theta = 0$ geometry but is isoenergetic with the *tbp* structure at the structurally unlikely geometry where $\theta = \pm 30°$
(i.e. all basal-axial angles are 120°). Hence, in the absence of *s* orbital interaction the
tbp geometry will always be the more stable of the two.

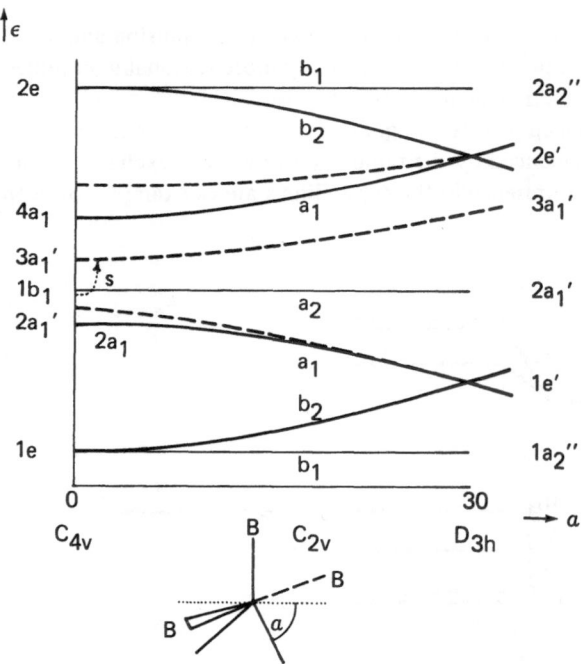

Fig. 16. Trigonal bipyramidal square pyramidal correlation diagram showing effect of *s* orbital
interaction. (At the C$_{4v}$ geometry $3\,a_1$ and $1\,b_1$ are equienergetic in the absence of *s* orbital
mixing.)

Inclusion of the s orbital on the central atom removes the equivalence of the pair of a_1 orbitals $3a_1$, $4a_1$ (in C_{4v}) and we may draw the correlation diagram of Fig. 16. At the D_{3h} geometry all the ligand-central atom s orbital interaction is concentrated in the $3a_1$ orbital. (The other p orbital manifold a_1' orbital of the tbp structure correlates with the $1b_1$ orbital of the square pyramidal geometry via an a_2 label in the C_{2v} intermediate geometry. Its s orbital interaction will therefore be zero throughout the domain of Fig. 16.) On distortion to C_{4v}, this $3a_1$ orbital becomes stabilised as the latent s orbital destabilisation energy is shared with $2a_1$ and $4a_1$ orbitals derived from $1e'$ and $2e'$ of the trigonal bipyramid. For systems with five valence pairs (all orbitals up to $2a_1$ doubly occupied) as in PF_5 the tbp geometry may be readily seen to be more stable than the C_{4v} geometry. Fig. 16 indicates that it is inclusion of s orbital interaction that strongly favours the tbp geometry. From this discussion we should be able to see the factors which bring the stabilisation energy of the spy geometry closer to that for the tbp geometry for the five valence pair case. Firstly, the larger the droop angle of C_{4v} the closer the two quartic terms. At the isoenergetic point of $\theta = 30$ however we expect strong steric repulsions to exist between the equatorial ligands (cf. the C_{4v} geometry for the AB_4 case). Secondly, a small amount of s orbital involvement would lead to a small s orbital destabilisation of $2a_1$. This would be the case where the ligands are not very electronegative compared to the central atom. This, however, does not seem to be a good explanation for the geometries of $InCl_5^{-2}$ or $Sb\Phi_5$ where the ligand electronegativity is not low. Both of these examples however are not true C_{4v} molecules but are significantly distorted to C_{2v}.

Inclusion of two more electrons leads to population of $3a_1'$ which is rapidly stabilised on distortion away from the tbp geometry. For molecules with this configuration e.g. BrF_5 and ClF_5 the square pyramidal geometry should be observed. Due to the slope of $3a_1$ on distortion negative droop angles should be favoured as is found to be the case.

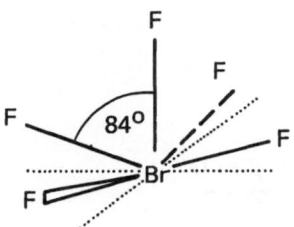

The VSEPR approach attributes this geometry to the electrostatic repulsion due to the sixth pair of electrons sticking out of the base of the square pyramid. Isoelectronic XeF_5^+ has a droop angle of $11°$ (axial-equatorial angle of $79°$). The present approach attributes the stability of the spy geometry to the fact that here the s orbital destabilisation is smallest and the observed bond angles due to the same molecular orbital result we have seen earlier in ClF_3 and SF_4.

7. AB₆ Systems and IF₇

Here we initially compare the stabilisation energies of the octahedral and trigonal prismatic structures. Molecular orbital diagrams for the two systems are shown in Fig. 17, 18 with the energy dependences calculated from Eq. 4. We may consider the trigonal prism to be constructed from two C_{3v} AB₃ units sharing an apex with a droop angle from planar within each unit of angle θ. On the basis of quadratic terms with three p manifold pairs or more the two structures are isoenergetic. Inclusion of quartic terms leads to identical values (of $-24\gamma_\sigma S_\sigma^4$) for the octahedron and trigonal prism

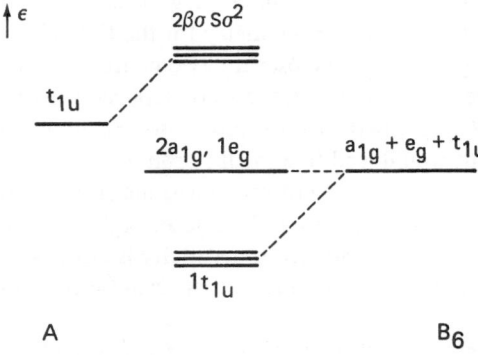

Fig. 17. Molecular orbital diagram for AB₆ octahedral system.

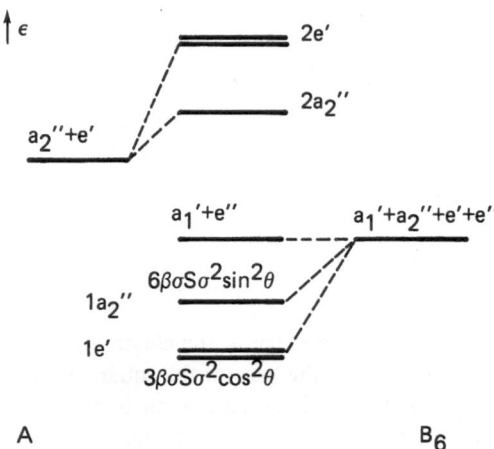

Fig. 18. Molecular orbital diagram for AB₆ trigonal prism system.

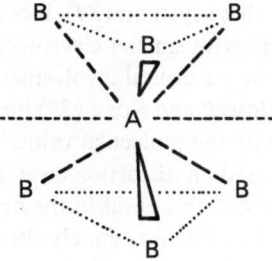

with an angle of $\theta = \sin^{-1} \sqrt{1/3}$ (i.e. the orthogonal structure). In the case of the trigonal prism non-bonded repulsions would be expected to increase θ from this value and thus increase the size of the quartic stabilisation energy term. Such a process would favour the octahedron, although electronically there is no reason to prefer this geometry. In Fig. 19 s orbital interactions are included and these obviously destabilise the relevant a_1' and a_{1g} orbitals of each structure by an equal amount since only one example of each exists in this orbital grouping. However, whereas the two geometries may be close in energy there is a barrier to their interconversion as can be readily seen in the following way from Fig. 19. At the D_{3h} and O_h geometries, the total s orbital interaction is contained in the $2a_{1g}$ and $2a_1'$ orbitals respectively. However, in the C_{3v} geometry intermediate between the two, the ligand orbitals transform as $2e + 2a_1$ and the destabilisation energy is now shared between the *three* a_1 orbitals as in Fig. 19. For SF_6 and SF_5Cl therefore (six valence pairs) the molecule is held in the octahedral configuration due to this 's orbital' barrier. Room temperature *nmr*

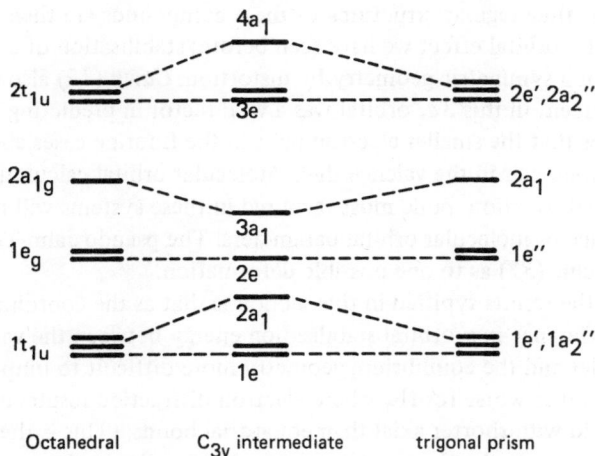

Fig. 19. Correlation diagram octahedral-trigonal prism for AB_6 molecules showing s orbital destabilisation of all a_1 orbitals derived from non a_1 representations of parent geometries.

data indicate no intramolecular rearrangement in these systems (*34*). More facile rearrangement would be expected to occur for systems with smaller electronegativity differences between central atom and ligands (since here *s* orbital involvement would be smaller) but there is no proven case. *Hoffmann, Howell* and *Rossi* (*35*) however conclude that SF_6 is held in the octahedral geometry by the molecular orbital equivalent of non-bonded repulsions which are rapidly increased on distortion away from the octahedral geometry. We shall have more to say about this point in the next section.

For XeF_6 however (with seven valence pairs) Fig. 19 immediately shows that with occupation of the $3a_1$ orbital that distortions away from octahedral may now be favoured. Whilst we have shown just one such distortion in Fig. 19 (via the trigonal twisting mechanism) in principle any distortion which leads to a decrease in *s* orbital destabilisation of the *p* manifold a_1 orbitals will be allowed. This will be one where an orbital of symmetry a_1 is created from amongst the *p* manifold bonding orbitals which do not transform as a_{1g} under the octahedral point group (as in Fig. 19). Whether the ocathedral geometry will be susceptible to such distortion will depend upon the balance between the gain in stabilisation due to this *s* orbital effect and changes in the magnitude of the quartic terms as the geometry changes. (With all bonding orbitals filled there will be no change in quadratic energy terms on changing the geometry.) Bearing this in mind we may readily see that systems with large electronegativity differences between central atom and ligand will (a) have larger *s* orbital involvement and (b) smaller values of $\gamma_\sigma S_\sigma^4$ than systems where the electronegativity difference is small. Thus such distortions are most likely to occur for molecules where a large electronegativity difference exists. This is found to be the case. The MX_6^{-2} (M = Se, Te; X = Cl, Br) systems have regular octahedral structures as does $SbCl_6^{-3}$ whereas IF_6^- and XeF_6 are definitely distorted. The explanation is not a good one to explain the distorted nature of $BiBr_6^{-3}$ however. (This however may be a crystal effect since $SbCl_6^{-3}$ (octahedral in solution) is distorted (*36*) in the crystal.) The distorted, rather than regular structures of these compounds are then due again to the same molecular orbital effect we have seen before; stabilisation of a highly unstable a_1 orbital of a symmetric geometry by distortion. *Gavin* (*12*) also noted that the *s* orbital involvement in this $3a_1$ orbital was a vital factor in predicting distorted XeF_6. VSEPR argues that the smaller electron pairs in the fluorine cases above allow more room for the lone pair in the valence shell. Molecular orbital calculations intent on discovering the deformation mode most favoured in these systems will be very sensitive to the choice of molecular orbital parameters. The pseudo Jahn-Teller argument gives a useful clue (*37*) as to one possible deformation.

One feature of the results typified in this section is that as the coordination number increases the difference in *p* orbital stabilisation energy between the various structures becomes smaller and the equilibrium geometry more difficult to pinpoint. This problem should be rather worse for IF_7 where electron diffraction results indicate a pentagonal bipyramid with shorter axial than equatorial bonds. (This is the reverse to that found for the *tbp* in the five coordinate case.) Using the methods described earlier the molecular orbital diagram for this geometry is shown in Fig. 20 and we may readily confirm the observed bond length difference using the method described

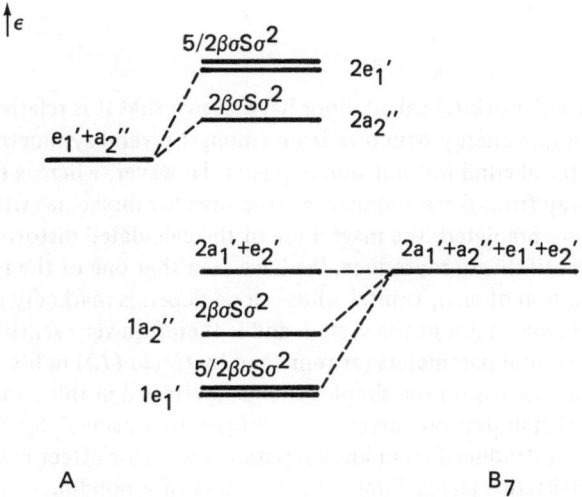

Fig. 20. Molecular orbital diagram for D_{5h} AB_7 molecule. (*s* orbital interaction neglected.)

above. Stabilisation energies of the two types of linkage are $2\beta_\sigma S_\sigma^2 - 4\gamma_\sigma S_\sigma^4$ per axial ligand and $2\beta_\sigma S_\sigma^2 - 5\gamma_\sigma S_\sigma^4$ per equatorial ligand.

Three possible geometries, each considered at various times for this molecule, are shown in Fig. 21. We do not wish to try to distinguish energetically between them. However, we do note that the D_{5h} geometry is the only one where no *s* orbital destabilisation occurs amongst the lowest occupied *p* manifold orbitals. In the C_{2v} point group $e_1' \rightarrow a_1 + b_1$ and thus one component of e_1' of Fig. 20 is destabilised when the symmetry is lowered. In the C_{3v} point group $a_2'' \rightarrow a_1$ which also then receives a severe destabilisation on moving from D_{5h}. On this basis the pentagonal bipyramid should be the most stable geometry for IF_7.

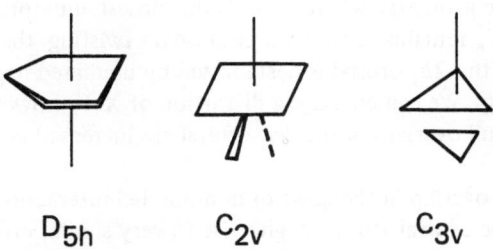

Fig. 21. Possible structures for seven coordinate molecules.

8. Discussion

Previous simple molecular orbital calculations have shown that it is relatively simple to pinpoint the minimum energy structure from amongst several symmetric geometries. e.g. CH_4 as tetrahedral but not square planar. However, whereas the direction of distortion away from these symmetric structures for molecules with more or less electrons is readily predicted, the magnitude of the calculated distortion is much more difficult to quantitatively reproduce. We have seen that one of the reasons for this lies in the population of an a_1 orbital whose slope depends markedly on the amount of s orbital involvement in the system and is therefore very sensitive to the choice of molecular orbital parameters (as remarked by *Gavin* (*12*) in his simple treatment of XeF_6). Thus whereas on the simple scheme developed in this paper, we could calculate the p orbital stabilisation energy of a configuration using $\beta_\sigma S_\sigma^2$ and $\gamma_\sigma S_\sigma^4$, inclusion of s orbitals introduced an unknown parameter whose effect could only readily be seen in qualitative terms. Similarly, the effect of π bonding — neglected here for the most part — can alter some of the finer points of the molecular geometry.

We noted in Section 4 that by introducing non-bonded interactions into the molecular orbital scheme we would also lose the non-bonding character of the a_1 non-bonding orbitals involved in the p orbital manifold in a similar sort of way to introduction of a central atom s orbital. We illustrate some cases in Fig. 22.

We saw above that BF_3 and NF_3 do not distort to a T shape geometry because the $2a_1$ orbital has an unfavourable slope. We may regard this unfavourable slope also as being due to an increase in non-bonded repulsions through the increased out-of-phase overlap in this doubly occupied molecular orbital (Fig. 22a). Similarly the favourable slope of $3a_1$ leads to a T-shaped ClF_3 molecule. This is consistent with the in-phase nature of this orbital which results in increased overlap on distortion (Fig. 22b). For the four coordinate examples, SF_4 may be considered to distort from square planar to disphenoidal due to the in-phase nature of the $3a_1$ orbital of Fig. 13 as shown in Fig. 22c, which leads to increased ligand-ligand overlap on bending. CF_4 however does not distort to square planar (Fig. 22d) because of the unfavourable change in ligand-ligand overlaps in $2a_1$ of Fig. 10. Similarly, PF_5 is D_{3h} because $2a_1$ of Fig. 16 favours the *tbp* geometry (Fig. 22e) but ClF_5 is C_{4v} because the largest non-bonded overlap occurs at the geometry where there is the closest equatorial-axial ligand approach (Fig. 22f). SF_6 remains octahedral because on twisting, the ligand ligand distance decreases and the $2a_1$ orbital is destabilised by increased negative ligand-ligand overlap (Fig. 22g). We can envisage a distortion of XeF_6 however from octahedral where ligand-ligand overlaps in the $3a_1$ orbital are increased compared to octahedral (Fig. 22h).

Thus both ligand-ligand overlap in the guise of non-bonded interactions and introduction of an s orbital on the central atom can give rise to very similar effects in the behaviour of the molecular orbital energies on distortion. Either may be the agent to remove the non-bonding nature of some of the a_1 orbitals of the molecule. It is thus very interesting to note that *Hoffmann* and *co-workers* (*35*) attributed the rigid structure of SF_6 to the 'molecular orbital equivalent of non-bonded interactions' whereas in this paper we have stressed the importance of central atom s involvement. Obviously, both will be important but in the examples we have chosen above the two effects reinforce rather than cancel each other.

Table 3 gives a summary of the effects determining molecular geometry in the series of molecules we have looked at illustrated by known fluoride examples.

Table 3. Summary of the Effects Determining Molecular Shapes as a Function of the Number of Ligands and Number of Valence Pairs of Electrons

Ligands	Nr. Electron Pairs					
	2	3	4	5	6	7
2	quadratic terms BeF_2	quartic terms BF_2	quartic terms OF_2	quadratic terms XeF_2		
3		quadratic terms BF_3	quartic terms NF_3	a_1 orbital changes ClF_3		
4			quartic terms CF_4	a_1 orbital changes SF_4	quadratic terms XeF_4	
5				quartic terms PF_5	a_1 orbital changes ClF_5	
6					quartic terms SF_6	a_1 orbital changes XeF_6
7					?	IF_7

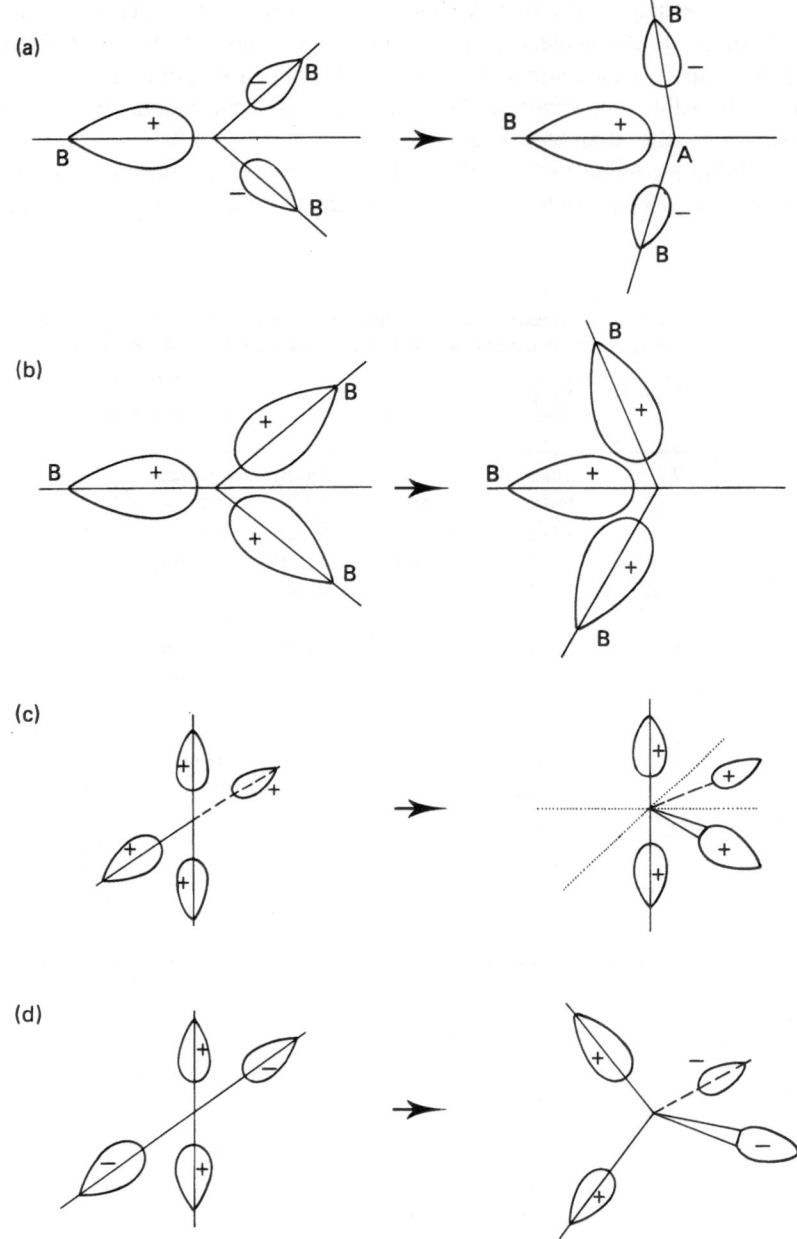

Fig. 22. Changes in the size and sign of non-bonded interactions on distortion AB$_3$ systems $2\,a_1$ (a) and $3\,a_1$ (b) orbitals; AB$_4$ systems $3\,a_1$ (c) and $2\,a_1$ (d) orbitals; AB$_5$ systems $2\,a_1$ (e) and $3\,a_1$ (f) orbitals; AB$_6$ systems $2\,a_1$ (g) and $3\,a_1$ (h) orbitals.

The Shapes of Main-Group Molecules

(e)

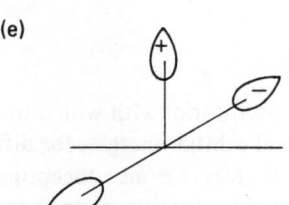 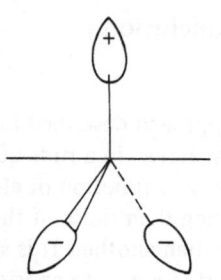

(f)

 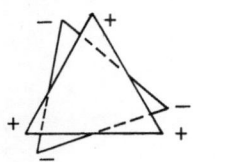 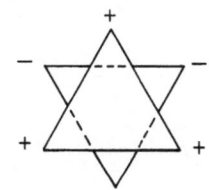

(g)

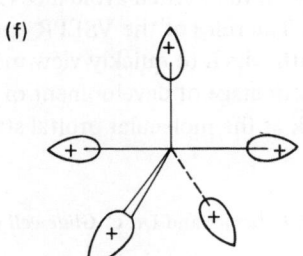 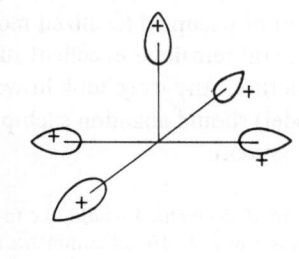

(h)

 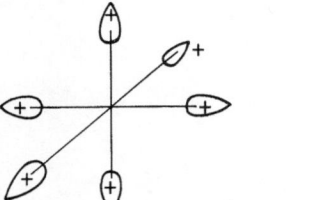 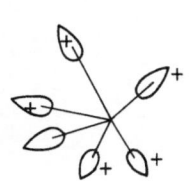

103

9. Conclusion

The approach described in this work provides a useful tool with which to approximately derive, in a rude quantitative fashion, total orbital energies for different geometries as a function of electron configuration. It gives reasons conceptually more satisfying than those of the VSEPR approach for the adoption of one geometry rather than another. It is similarly able to rationalise relative bond lengths and angles without recourse to considerations involving repulsions of pairs of electrons. Whereas it is relatively easy to put these molecular orbital ideas on a plausible semi-quantitative basis it is much more difficult to do the same with the mutual avoidance (Pauli avoidance) of occupied localized molecular orbitals. The rules of the VSEPR approach however still remain as excellent rules of thumb with which to quickly view molecular structure. Any close look however (at the present stage of development of the VSEPR model) should abandon such premises and look at the molecular orbital structure of the system.

Acknowledgement: I would like to thank Professor *J.J. Turner* and Dr. *C. Glidewell* who read the manuscript and offered comments and criticisms.

References

1. *Sidgwick, N. V., Powell, H. M.:* Proc. Roy. Soc., *A176*, 153 (1940).
2. *Gillespie, R. J., Nyholm, R. S.:* Quart. Rev., *11*, 339 (1957).
3. *Fowles, G. W. A.:* J. Chem. Educ., *34*, 187 (1957).
4. *Gillespie, R. J.:* J. Chem. Educ., *40*, 295 (1963); J. Chem. Soc. 4672, 4679 (1963); Can. J. Chem., *38*, 818 (1960); ibid, *39*, 318 (1961).
5. *Gillespie, R. J.:* Molecular Geometry, London: Van Nostrand-Rheinhold (1972).
6. *Walsh, A. D.:* J. Chem. Soc., 2260, 2266, 2288, 2296, 2301 (1953).
7. *Schnuelle, G. W., Parr, R. G.:* J. Amer. Chem. Soc. *94*, 8974 (1972).
8. *Bartell, L. S.:* J. Chem. Educ., *45*, 754 (1968).
9. *Pearson, R. G.:* J. Amer. Chem. Soc., *91*, 1252, 4947 (1969).
10. *Bartell, L. S.:* J. Chem. Phys., *32*, 827 (1960).
11. *Glidewell, C.:* Inorg. Chim. Acta, *12*, 219 (1975).
12. *Gavin, R. M.:* J. Chem. Educ., *46*, 413 (1969).
13. *Hoffmann, R., Howell, J. M., Muetterties, E. L.:* J. Amer. Chem. Soc., *94*, 3047 (1972) and references therein.
14. *Gimarc, B. M.:* J. Amer. Chem. Soc., *92*, 266 (1970).
15. *Burdett, J. K.:* J. Chem. Soc. (Faraday II), *70*, 1599 (1974).
16. *Elian, M., Hoffmann, R.:* Inorg. Chem., *14*, 1058 (1975).
17. *Burdett, J. K.:* Inorg. Chem., *14*, 375, (1975).
18. *Larsen, E., La Mar, G. N.:* J. Chem. Educ., *51*, 633 (1974).
19. *Jørgensen, C. K., Pappalardo, R., Schmidke, H.-H.:* J. Chem. Phys., *39*, 1422 (1963).
20. *Schaffer, C. E., Jørgensen, C. K.:* Mol. Phys., *9*, 401 (1965).
21. *Schaffer, C. E.:* Structure and Bonding, *5*, 68 (1968).
22. We may readily show this by differentiation of the term in parentheses with respect to α and setting it equal to zero.
23. In fact, the total s orbital stabilisation energy *will* change as the geometry changes since the β term in Eq. 1 contains the inverse of the energy separation between (in this case) s orbital and ligand a_1 (σ) − central atom orbitals. The energies of the latter will change on distortion but this effect will be smaller.
24. *Current, J. H., Burdett, J. K.:* J. Phys. Chem., *73*, 3505 (1969).
25. *Karplus, M., Fraenkel, G. K.:* J. Chem. Phys., *35*, 1312 (1961); *Fessender, R. W., Schuler, R. H.:* J. Chem. Phys., *39*, 2147 (1963).
26. *Fessender, R. W., Schuler, R. H.:* J. Chem. Phys., *43*, 2704, (1965).
27. *Maass, G., Maltsev, A. K., Margrave, J. L.:* J. Inorg. Nucl. Chem., *35*, 1945 (1973).
28. *Mulliken, R. S.:* J. Phys. Chem., *41*, 318 (1937); *56*, 295 (1952).
29. *Bent, H. A.:* J. Chem. Educ., *45*, 768 (1968); *40*, 446, 523 (1963).
30. *Musher, J. I.:* Angew. Chem. (Int. Ed.), *8*, 54 (1969).
31. See for example the diagrams in Ref. 12.
32. See for example *F. A. Cotton and G. Wilkinson* 'Advanced Inorganic Chemistry', 3rd edn., London (1972).
33. *Burdett, J. K.:* Inorg. Chem., *15*, 212 (1976).
34. *Merrill, C. I., Williamson, S. M., Cady, G. H,, Eggers, D. F.:* Inorg. Chem., *1*, 215 (1962).
35. *Hoffmann, R., Howell, J. M., Rossi, A. R.:* Inorg. Chem., in press (1976)
36. *Martineau, E., Milne, J. B.:* J. Chem. Soc. (A), 2971, (1970).
37. *Bartell, L. S., Gavin, R. M.:* J. Chem. Phys., *48*, 2466 (1968).

Author-Index Volume 1 — 31

CH. K. JOERGENSEN

Oxidation Numbers and Oxidation States

1969. VII, 291 pages

(Molekülverbindungen und Koordinationsverbindungen in Einzeldarstellungen)

Contents: Formal Oxidation Numbers. Configuration in Atomic Spectroscopy. Characteristics of Transition Group Ions. Internal Transitions in Partly Filled Shells. Inter-Shell Transitions. Electron Transfer Spectra and Collectively Oxidized Ligands. Oxidation States in Metals and Black Semi-Conductors. Closed-Shell Systems, Hydrides and Back-Bonding. Homopolar Bonds and Catenation. Quanticule Oxidation States. Taxological Quantum Chemistry.

Electrons in Fluids

The Nature of Metal-Ammonia Solutions.
Editors: J. Jortner, N. R. Kestner.
1973. 271 figures, 59 tables. XII, 439 pages

This full and up-to-date account of the chemical and physical properties of electrons in polar, nonpolar, and dense fluids includes contributions from both theoretical and experimental chemists and physicists, thus clearly indicating the interdisciplinary nature of this field.

S. P. SINHA

Europium

1967. 23 figures. VIII, 164 pages

(Anorganische und allgemeine Chemie in Einzeldarstellungen, Band 8).

Springer Verlag
Berlin
Heidelberg
New York

Contents: Methods of Separation of Individual Rare Earth Elements. Preparation and Properties of Europium. Alloys and Intermetallic Compounds of Europium. Compounds of Europium. Spectroscopic Properties of Europium. Possible Uses and Applications of Europium.

Topics in Current Chemistry

Fortschritte der chemischen Forschung
Managing Editor: F. L. Boschke

Vol. 60
Structure of Liquids

1975. 88 figures, 38 tables. IV, 205 pages.

Contents/Information: P. Schuster, W. Jakubetz, W. Marius: Molecular Models for the Solvation of Small Ions and Polar Molecules.
Recent developments in the theory of interaction between ions or polar molecules and their nearest neighbors in the solution are summarized. New experimental data are presented. (268 references)
S. A. Rice: Conjectures on the Structure of Amorphous Solid and Liquid Water.
Enormous effort has been invested in the experimental determination of the properties of water and water based solutions. This article summarizes the results of some recent studies of amourphous solid water, suggests a correlation of its properties with those of the liquid, and drawing on the common features of a variety of models, projects a conjecture concerning the structure of water. (94 references)

Vol. 64
Inorganic Biochemistry

1976. 85 figures. IV, 225 pages.

Contents/Information: E. T. Degens: Molecular Mechanisms on Carbonate, Phosphate, and Silica Deposition in the Living Cell.
Focal point is the molecular interaction between metal ions and skeletal tissues in "healthy" cells. Special attention is given phenomena leading to the nucleation of a crystal seed and its subsequent oriented growth in the direction of an organized structure. (585 references)
W. A. P. Luck: Water in Biologic Systems.
During the last years biologists and physicians have recognized more and more the important role of water, but there are very few places where one can specify the exact role water is playing. (303 references)
D. D. Perrin: Inorganic Medicinal Chemistry.
Only limited attempts appear to have hitherto been made to bring together what might be described as inorganic medicinal chemistry. (192 references)

Vol. 65
Theoretical Inorganic Chemistry II

1976. 47 figures, 44 tables. IV, 153 pages.

Contents/Information: K. Bernauer: Diastereoisomerism and Diastereoselectivity in Metal Complexes.
Diastereoselectivity of a single ligand molecule has as yet not been reviewed. Even though some examples of conformational analysis are known and predictions of selectivity can be made, the orientation of a secondary ligand in a chiral mixed ligand complex is still based on empirical data. (134 references)
M. S. Wrighton: Mechanistic Aspects of the Photochemical Reactions of Coordination Compounds.
Increasing interest in the photochemistry of transition metal containing molecules prompts this survey of the important developments in understanding the chemical transformations resulting from electronic excitation of such molecules. (196 references)
A. Albini, H. Kisch: Complexation and Activation of Diazenes and Diazo Compounds by Transition Metals.
The interest in the coordinating properties of diazenes stems from the probable intermediacy of diazene complexes in biological nitrogen-fixation as well as from the nonenzymatic conversion of coordinated dinitrogen into diazene derivatives. (119 references)

Springer-Verlag Berlin Heidelberg New York